Content-based Microscopic Image Analysis

DISSERTATION

zur Erlangung des Grades eines Doktors
der Ingenieurwissenschaften

vorgelegt von

M. Sc. Chen Li

geb. am 22.04.1985 in China

eingereicht bei der Naturwissenschaftlich-Technischen Fakultät

der Universität Siegen

Siegen 2015

stand: November 2015

Studien zur Mustererkennung

herausgegeben von:

Prof. Dr.-Ing. Heinrich Niemann
Prof. Dr.-Ing. Elmar Nöth

Bibliografische Information der Deutschen Nationalbibliothek

Die Deutsche Nationalbibliothek verzeichnet diese Publikation in der
Deutschen Nationalbibliografie; detaillierte bibliografische Daten sind
im Internet über http://dnb.d-nb.de abrufbar.

ISBN 978-3-8325-4253-5
ISSN 1617-0695

Logos Verlag Berlin GmbH
Comeniushof
Gubener Str. 47
10243 Berlin
Tel.: +49 030 42 85 10 90
Fax: +49 030 42 85 10 92
INTERNET: http://www.logos-verlag.de

Gutachter der Dissertation:

1. Prof. Dr.-Ing. Marcin Grzegorzek

2. Prof. Dr.-Ing. Klaus-Dieter Kuhnert

Tag der mündlichen Prüfung:

16. Februar 2016.

Acknowledgements

First of all, I would like to express my utmost gratitude to my first supervisor, Prof. Dr.-Ing. Marcin Grzegorzek, for his supervision, advice, and guidance on this research and encouragement and support throughout the time of my study at the University of Siegen. His broad knowledge in pattern recognition, infinite wisdom, modesty and spirit inspired and enriched my growth as a student and a researcher that I want to be, and enlightened me very much to find solutions whenever I encountered problems during my PhD studies. Then, I wish to give great thanks to my second supervisor Prof. Dr.-Ing. Klaus-Dieter Kuhnert for his huge help and support regarding my academic development. And I also thank Prof. Dr. Roland Wismüller and Prof. Dr.-Ing. Madjid Fathi a lot for their positive evaluation to my PhD work.

I would like to give my thanks and appreciation to my project leader, Dr. Kimiaki Shirahama, for his consistent help, invaluable academic guidance and attention during the entire process of this work. His commitment, scientific knowledge, and continuous support during the past years were crucial not only for the accomplishment of this work, but also for the expansion of my scientific knowledge and my growing interest in scientific studies. For me, he is the Mr. Fujino to Mr. Lu Xun.

I want to thank my parents for always encouraging me to pursue whatever interest I might have fancied at the moment. I am grateful that I have had the opportunity to find my own path in life and that I have always felt your support. I am also grateful for the support I have received from my parents-in-law, not only during my PhD studies. Furthermore, long live all my grandparents and relatives!

The gratitude I feel towards my wife Ning Xu is beyond expression. You are the best! Without you, I probably would have never even started this journey, and without you, I would not have finalised it. Thank you for all the joy and support you have given me. Thank you for the many memorable times staying in Germany with you.

Thanks to all of my friends, colleagues and coauthors in Germany whom I had a wonderful time with, and the friends in China who share my failures and successes. In particular, I thank Prof. Dr. Kuniaki Uehara, Prof. Dr. Oliver Deussen and Prof. Dr. Beihai Zhou for their great academic support; I thank Prof. Dr. Tao Jiang, Dr. Minmin Shen and Dr. Fangshu Ma for their excellent long-term cooperation; I thank Dr.-Ing. Joanna Czajkowska and Dipl.-Inform. Christian Feinen for their friendly tutoring at the beginning of my PhD study; I thank Dr. Peter Volz, M.Sc. Zeyd Boukhers, M.Sc. Lukas Köping, Ms. Kristin Klaas, Mr. Florian Schmidt and Mr. Oliver Tiebe for their helpful technical support in many aspects; I also thank everyone else, thank you very much for your help of all kinds.

I am thankful for the financial support I received from the China Scholarship Council, and the graduation aid programme of the DAAD at the University of Siegen.

Abstract

In this dissertation, novel *Content-based Microscopic Image Analysis* (CBMIA) methods, including *Weakly Supervised Learning* (WSL), are proposed to aid biological studies. In a CBMIA task, noisy image, image rotation, and object recognition problems need to be addressed. To this end, the first approach is a general supervised learning method, which consists of image segmentation, shape feature extraction, classification, and feature fusion, leading to a semi-automatic approach. In contrast, the second approach is a WSL method, which contains *Sparse Coding* (SC) feature extraction, classification, and feature fusion, leading to a full-automatic approach. In this WSL approach, the problems of noisy image and object recognition are jointly resolved by a region-based classifier, and the image rotation problem is figured out through SC features. To demonstrate the usefulness and potential of the proposed methods, experiments are implemented on different practical biological tasks, including environmental microorganism classification, stem cell analysis, and insect tracking.

This dissertation is structured as follows: Chapter 1 first prefaces the whole thesis with a brief introduction. Then, Chapter 2 reviews related CBMIA techniques and applications. Next, a semi-automatic microscopic image classification system is proposed in Chapter 3, using strongly supervised learning methods. In contrast, full-automatic WSL approaches are introduced in Chapter 4. After the aforementioned supervised learning methods, an unsupervised learning approach is developed in Chapter 5 to solve problems regarding microscopic image clustering. Following that, Chapter 6 introduces a CBMIA framework for the task of multi-object tracking. Then, the proposed CBMIA methods are tested on three microbiological tasks to prove their usefulness and effectiveness in Chapter 7, including environmental microorganism classification, stem cell clustering and insect tracking. Finally, Chapter 8 completes the whole dissertation with an overall conclusion of the current work and a brief plan for the future research.

Zusammenfassung

In der vorliegenden Dissertation werden neue Methoden zur inhaltsbasierten, mikroskopischen Bildanalyse (CBMIA) vorgestellt. Die Ansätze basieren dabei unter anderem auf Techniken aus dem schwach überwachten Lernen und finden vor allem Anwendung in biologischen Aufgabenbereichen. In der inhaltsbasierten, mikroskopischen Bildanalyse müssen Probleme wie verrauschte Bilder, Bildrotation und Objekterkennung adressiert werden. Hierfür stützt sich der erste vorgestellte Ansatz auf eine semi-automatisierte Technik, die eine allgemeine, überwachte Lernmethode darstellt. Der Ansatz beruht auf der Segmentierung von Bildern, auf von der Objektform abhängigen Merkmalen und auf der Klassifikation und Fusion von Merkmalen. Im Gegensatz dazu handelt es sich beim zweiten Ansatz um eine schwach überwachte Lernmethode. Die Merkmalsextraktion erfolgt hierbei durch Sparse Coding (SC). Zusätzlich beinhaltet sie eine Klassifikation und Merkmalsfusion, was letztlich zu einem voll-automatisierten System führt. Im schwach überwachten Lernansatz werden die zwei Probleme, Bildrauschen und Objekterkennung, gemeinsam mittels eines auf Bildregionen basierendem Klassifikators gelöst, während das Problem der Bildrotation mittels SC angegangen wird. Die Nützlichkeit und Effektivität der vorgestellten Methoden wird anhand verschiedener Aufgaben aus dem Bereich der Biologie aufgezeigt. Diese Aufgaben beinhalten die Klassifikation von Mikroorganismen, Stammzellenanalyse und die automatisierte Verfolgung von Insekten.

Die Dissertation ist wie folgt aufgebaut: Kapitel 1 führt den Leser in einer kurzen Einleitung in das Thema ein. In Kapitel 2 folgt eine Aufarbeitung von verwandten Techniken aus dem Bereich der inhaltsbasierten, mikroskopischen Bildanalyse und seiner Anwendungsbereiche. Anschließend wird in Kapitel 3 ein semi-automatisiertes System zur Klassifizierung von Mikroskop-Bildern vorgestellt, das auf Methoden des stark überwachten Lernens basiert. Im Gegensatz dazu werden in Kapitel 4 voll-automatisierte, schwach-überwachte Lernmethoden eingeführt. Nach den gerade erwähnten überwachten Lernmethoden wird in Kapitel 5 ein unüberwachter Lernansatz entwickelt, um Probleme bezüglich der mikroskopischen Bildgruppierung zu lösen. Kapitel 6 führt ein CBMI basiertes Framework zur Multi-Objekt-Verfolgung ein. Im Anschluss werden in Kapitel 7 die vorgestellten Ansätze anhand von drei Aufgaben aus der Mikrobiologie auf ihre Effektivität und Nützlichkeit getestet. Die Aufgaben umfassen die Klassifikation von Mikroorganismen, die Gruppierung von Stammzellen und das automatisierte Verfolgen von Insekten. Die Dissertation wird in Kapitel 8 durch eine Zusammenfassung der Arbeit und Aufzeigen zukünftiger Forschungsvorhaben abgeschlossen.

Contents

List of Figures

List of Tables

Chapter 1

Introduction

Recently, microscopic images are analysed to help people to learn knowledge and understand essence in microcosmos. Especially, they play a very important role in biological fields, including microbiology and medical biology. Traditionally, microscopic images of biology are analysed by manual observation, where a lot of labour and time cost is needed. In order to increase efficiency of the analysis work, computer aided biology using *Content-based Microscopic Image Analysis* (CBMIA) approaches are developed and applied. In this chapter, fundamental concept of CBMIA is first introduced in Section 1.1. Then motivation and contribution of this dissertation is stated in Section 1.2 and Section 1.3, respectively. Finally, an overview of the whole work is provided in Section 1.4.

1.1   Fundamental Concept of Content-based Microscopic Image Analysis

Content-based Microscopic Image Analysis (CBMIA) is a branch of *Content-based Image Analysis* (CBIA). In generalized CBIA, an image is analysed by perceptual properties of itself rather than its metadata. Here, 'content' means all information that is able to be extracted automatically from the image itself, e.g. colours, textures and shapes. And 'metadata' means the individual information which describes the 'contents' of the image, e.g. tags, labels and keywords. 'Image analysis' is an approach which extracts meaningful information from images and represents them by numerical feature vectors [Sme+00; SB10]. Especially, CBMIA concentrates on the information extraction of 'content' of microscopic images.

Because CBMIA systems are usually semi- or full-automatic, they are effective and can save a lot of human resource. Furthermore, because CBMIA approaches only need some cheap equipment, like microscope and computer, the above analysis work can reduce a lot of financial investment. Hence, CBMIA can help people to obtain useful microcosmic information effectively, and it is widely used in many scientific and industrial fields [Orl+07], for example:

- Micro-operation [WLZ03; Son+06]:
  Nowadays, micro-operation are more and more used in minimally invasive surgeries, precise instrument troubleshooting and electric circuit maintenance. In micro-operation as shown in Figure 1.1(a), CBMIA is used as a fundamental technique to handle the accurate operating process.

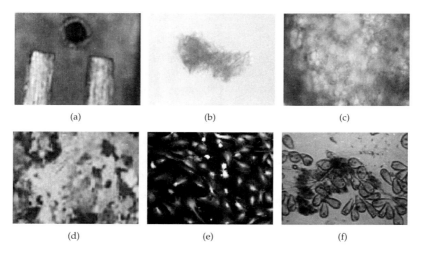

(a)   (b)   (c)

(d)   (e)   (f)

Figure 1.1: Examples of CBMIA domains. (a) Micro-operation [Son+06]: Two joy sticks are used to handle a micro-component. (b) Soil structure analysis: A piece of soil shows a typical granular structure under a microscope. (c) Plant tissue analysis: Structure of cell wall of a dry poplar leaf is observed. (d) Rock analysis [LM14]: The microscopic structure of a rock is revealed, where different elements show different colours. (e) Cell tissue analysis: Stem cells that are in a migration are recorded by microscopic images. (f) Microorganism analysis: *Epistylis* in activated sludge is checked to evaluate the pollution level of the environment.

- Soil Structure Analysis [RVB84; PV02]:
  Currently, CBMIA methods are effectively applied to analyse soil structure (see Figure 1.1(b)), which is a very important property relating to soil quality and agricultural production.

- Plant Tissue Analysis [SF06; SPC13]:
  The analysis of plant tissue is very important to keep ecological balance and
  diversity. CBMIA can aid this work in an effective and rapid way by analysing
  plant tissue images as shown in Figure 1.1(c).

- Rock, Metallographic and Mineral Analysis [LM14; Gee+07]:
  At present, material analysis is used in many industrial fields, just like quarrying,
  metallurgy and mining, where the rock, metallographic and mineral analysis
  is applied pervasively. To increase effectiveness of the analysis work, CBMIA
  approaches are used, e.g. in Figure 1.1(d).

- Cell Tissue and Biological Analysis [KR08; Wil12]:
  The analysis of cell tissue and biology is useful for medical treatment, agriculture, industry and environmental protection (see Figure 1.1(e) and 1.1(f)).

To achieve various application tasks as mentioned above, CBMIA needs to address different problems that occur in the analysis processes [VB11; Ban13; SHM11;
WMC08], for instance:

- High Noisy Image:
  As shown in Figure 1.2(a), some wicked microscopic conditions like low light or
  high impurity can lead to a high noisy image problem, which influences feature
  extraction seriously.

- Image Over-segmentation:
  In CBIMA, some tiny objects in a microscopic image often contain very important information, as shown in Figure 1.2(b). Because these objects are very
  thin and small to disconnect, an over-segmentation problem usually happens,
  resulting in a deficiency of useful information.

- Microscope Illuminant:
  Because in microscopic images the colours of objects are mainly decided by light
  sources of microscopes, the same object shows different colour characteristics
  based on different illuminations (see Figure 1.2(c)). This microscope illuminant
  problem leads to a troublesome case, where colour features lose effectiveness in
  some microscopic images.

- Image Rotation:
  Directionality of a microscopic image is different from that of a daily life image.
  The former does not have a specified orientation (or a positive direction) as
  shown in Figure 1.2(a), but the latter has a regular direction of gravity (e.g. a
  standing person in an image is always "head above, feet bottom"). This image

rotation problem makes some rotation sensitive features lose usefulness for most microscopic images.

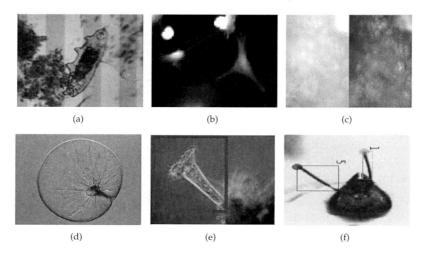

(a)   (b)   (c)

(d)   (e)   (f)

Figure 1.2: Examples of problems addressed by CBMIA. (a) An undirected image of *Dicranophorus* with a lot of noise and impurities. (b) A stem cell with very thin cytoplasm that is easy to over-segment. (c) The same plant leaf under natural (left) and artificial (right) light sources, which shows totally different colours in pink and orange. (d) An image of transparent *Noctiluca* which has no obvious texture and colour characteristics. (e) An image of detected *Stentor* (in the red bounding box), excluding impurities at the bottom right corner. (f) Two fast moving antennas (in the red and blue bounding boxes) of a honey bee are tracked.

- Opaque Object Representation:
  In microscopic images, description of an opaque object (e.g. in Figure 1.2(a)) is a general problem to be addressed, where colour, shape and texture features are often used.

- Transparent Object Representation:
  In contrast to opaque objects, transparent object description is a more difficult problem in CBMIA. As shown in Figure 1.2(d), the inside and outside of this transparent microorganism have consistent colour, so colour and texture features are nearly not useful to distinguish this microorganism itself from its background.

- Object Detection or Localisation:
  How to detect or localise an interesting object is an important problem addressed by CBMIA, which is helpful to discover unconspicuous objects, mark positions and exclude (or decrease) noise around objects (as shown in Figure 1.2(e)).

- Small Dataset:
  In real world, environmental conditions just like temperature, salinity and atmospheric pressure, are changing constantly. These unstable conditions influence sampling work of some sensitive objectives very much, for example: Figure 1.2(d) shows *Noctiluca* which is a usual index of water quality. It lives in seawater, so when people want to test the water quality in an estuary of a freshwater river where the salinity is low, it is hard to obtain a large number of its samples, leading to a small dataset problem. This problem limits the effectiveness of many classifiers (e.g. naive Bayesian classifier) during their inadequate training processes.

- Object Tracking:
  For high-speed moving objects in a video, e.g. antennas of a bee shown in Figure 1.2(f), how to track the objects becomes a pivotal problem in CBMIA tasks. Without solving this problem, image processing, feature extraction or classification of the objects are difficult to implement.

- Object Classification:
  To identify categories of different objects is a fundamental problem addressed by CBMIA. To solve this classification problem is the final goal of many CBMIA algorithms and systems.

- Microscopic Image Retrieval:
  Content-based Microscopic Image Retrieval (CBMIR) is an important application domain of CBMIA methods. Given a query image, to search similar images from a dataset is a basic problem which is addressed in practical scientific and industrial information systems.

In order to solve the problems above, CBMIA uses different techniques to handle them, respectively. The used techniques include general CBIA methods for daily life images [DLW05] and special methodology for microscopic images [VB11], for example:

- Image Denoising:
  To solve high-noisy problem of a microscopic image, different methods are used [Ban13]. For example, filtering is an often applied denoising process, which can remove some undesired components from an image.

- Image Segmentation:
  For obtaining high quality image segmentation results without suffering from the over-segmentation problem, all the existing CBMIA segmentation methods make great efforts to separate tiny objects as carefully as possible to keep more details of the object [SHM11]. For instance, edge detection based segmentation is a kind of useful method to obtain shapes of objects in an image.

- Feature Extraction:
  Because shape features can ignore the optical impact in microscopic images, they are the mostly used object descriptors in CBMIA work [WMC08]. For example, geometrical measures are very normally used shape features.

- Classifier Design:
  To classify objects in microscopic images, classifiers are designed to identify them into different categories [Sim+11]. For instance, linear classifier is a type of effective classifying algorithm.

- Object Tracking:
  In order to get to know the biological behaviours, object tracking is applied more and more usually in CBMIA work [RE04]. For example, tracking of single and multiple objects is developed and widely applied in recent years.

- Image Retrieval:
  Using a query microscopic image to search for similar images is the goal of CBMIR, where different approaches are used [EG99; AG12]. Comparing to the traditional text-based image retrieval method, CBMIR methods do not need any prior information of the image.

Figure 1.3 shows the framework of a general CBMIA system. As shown from the relational arrows, all stages in the CBMIA approach form an organic entity, where each step is not independent. These stages are interrelated and useful for improving overall performance of the whole system [TK09].

1.2   Motivation of the Present Work

The research motivations of the present work are introduced as follows:

- Environmental Microorganism Classification:
  Due to industrialisation and urbanisation, more and more pollutants like waste-gas and water are discharged into human living environments. These pollutants make people's surroundings dirty and lead to a lot of serious diseases, such as

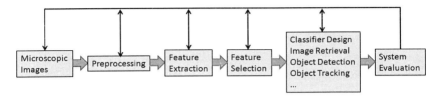

Figure 1.3: The general working flow of a CBMIA system.

malformations and cancers. To resolve such pollutants, one of the most natural and harmless approaches is to use *Environmental Microorganisms* (EMs), which are microscopic organisms living in natural environments (e.g. forests, rivers, mountains and oceans), and artificial environments (e.g. gardens, fish ponds and farmlands). Although people's usual impression on microorganisms may be harmful, EMs are useful for cleaning environments. Figure 1.4 shows some examples where EMs play important roles in environmental quality assessment and treatment. *Epistylis* in Figure 1.4(a) is a useful indicator for assessing the pollution level of activated sludge, because the amount of *Epistylis* increases when the sludge contains more organic waste. *Rotifera* in Figure 1.4(b), living in fresh water, is sensitive to salts and can be used to evaluate the quality of fresh water. When fresh water is polluted by inorganic salts, a large number of *Rotifera* disappears. Because *Vorticella* shown in Figure 1.4(c) can digest pollutants in wastewater very fast, it can be used to clean water in many wastewater plants. However, there are uncountable EMs on the earth, and people only have very limited knowledge about them. Thus, great breakthroughs are expected from environmental microbiology, which is the discipline to investigate functions, habits and characteristics of EMs [PGG14].

One of the most fundamental issues in environmental microbiology is 'EM classification' which classifies EMs into different categories. This is essential for researchers to get to know different operating conditions [Fri+00], habits and characteristics of different EMs [Lee+04]. For example, by analysing the relation between categories of EMs and environments, researchers can find that some EMs need warmer living environments, while others prefer cooler ones. This reveals that the temperature is an important operating condition for EMs. The above analysis is also valuable to find the relative relation among EMs, because EMs that are relatives favour the same living conditions and have similar habits. Beginning with such EMs, researchers can explore their difference in reproductive capacities. Thereby, among EMs that are relatives, researchers can find the ones that procreate faster and digest pollutants more quickly. In addition,

Figure 1.4: Examples of EMs in their corresponding environments. (a) *Epistylis* in sludge. (b) *Rotifera* in a river. (c) *Vorticella* in a wastewater plant.

when characteristics of various EMs are known, their classification can be used for assessing and treating environments. These are traditionally conducted by physical methods like filtration and sediment, and chemical methods like using chemical reagent. However, physical methods using high precision equipment require huge monetary cost. For chemical methods, when chemical reagents exceed the necessary quantity, they become a new chemical pollution. Compared to physical and chemical methods, using EMs is not only an effective approach which needs less economic cost, but a truly harmless approach which does not create any additional artificial pollution [SGA95; MC+01].

However, EMs are tiny and invisible to the naked eyes. Because their size varies usually between 0.1 to 100 μm , they can only be observed under a microscope. There are two traditional methods to recognise such EMs. The first is the 'morphological method' where an EM is observed under a microscope and recognised manually based on its shape [PGG14]. This process costs less fund and time, but the training process for a skillful operator is very time-consuming. Furthermore, even very experienced operators are unable to distinguish thousands of EMs without referring to literature. The second way is the 'molecular biological method' which distinguishes EMs by Deoxyribonucleic Acid (DNA) or Ribonucleic Acid (RNA) [GSL91; Ber+95]. Although this is an accurate approach, it is time-consuming and requires very expensive equipment. In order to solve problems of the above mentioned traditional methods, two efficient and practical EM classification systems based on CBMIA are developed to aid environmental microbiology. The first system uses a semi-automatic method, which is a very practical approach. The second system works in a full-automatic way, where the system becomes more intelligent.

- Object Structure Description:
  The structure of an object is an effective criterion for distinguishing its category. For example, *Arcella* in Figure 1.5(a) and *Actinophrys* in Figure 1.5(b) have similar sizes and circular shapes, but the latter is characterised by many pseudopodia and shows a radiate structure. Hence, *Internal Structure Histogram* (ISH) is proposed, which can effectively describe the differences of various structures.

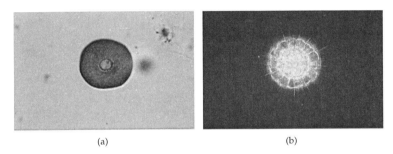

(a)   (b)

Figure 1.5: An example of different structures of objects. (a) A rounded structure of *Arcella*. (b) A radiate structure of *Actinophrys*.

- Image Segmentation:
  Image segmentation is an important preprocessing for shape feature extraction, where segmented images can support more exact shape information than original images. To this end, multiple segmentation methods are tested, and finally *Sobel Edge Detection* based approaches obtain the most robust results.

- Object Tracking:
  To improve the object tracking algorithms, an association based tracking approach is proposed to track multiple objects. Especially, videos of moving honey bees are used to test the efficiency and effectiveness of the method.

1.3   Contribution of the Present Work

This work is an interdisciplinary research between microscopy [MVD10] and computer science. Especially, it concentrates on the analysis of biological microscopic images using CBMIA methodologies. There are two main goals of the present work: The first is to explore effective CBMIA solutions for practical microscopic image analysis tasks. For this goal, different practical requirements of scientists are proposed by their professional domain knowledge. Based on their requests, suitable CBMIA methods are developed to solve various problems. This means, the developed CBMIA methods should consider problems in the real world, but are not only useful in a computer laboratory. The second is to improve the developed CBMIA approaches to be more intelligent and automatic. To this end, the semantic information of microscopic images is more deeply considered. By the semantic information, a machine is trained to analyse an image like a human being who recognises an object as an entirety.

To address the two goals mentioned above, the following respects are newly proposed in the present work:

- Environmental Microorganism Classification:
  EMs, such as *Dicranophorus* in Figure 1.2(a), are very tiny living beings in human surrounding environments, and can decompose pollutants as their nutrition source. So, they are important for environmental assessment and establishing a sustainable ecosystem. Hence, to get to know more knowledge of them, the identification of EM categories, the EM classification, plays a fundamental role. To this end, two novel EM classification approaches are proposed by processing images captured by microscopes. The first is a semi-automatic method using image segmentation, shape features, *Strongly Supervised Leaning* (SSL) and classifier fusion. In contrast, the second is a full-automatic method applying *Sparse Coding* (SC) features and *Weakly Supervised Learning* (WSL) methods. Experimental results on real-world images show the effectiveness of the proposed EM classification systems.

- Internal Structure Histogram:
  ISH is a contour-based shape descriptor, which captures the structure of an object using angles defined on its contour. This feature shows robust descriptive power in image classification and retrieval tasks on both microscopic and normal images.

- Sobel Edge Detection Based Image Segmentation:
  Because Sobel edge detection has a high robustness for noisy images, it is selected for image segmentation tasks in this CBMIA work. Based on Sobel edge

detection, two image segmentation methods are developed. The first is a semi-automatic approach, which is able to localise an object with a high accuracy. The second is a full-automatic method named as 'double-stage segmentation'. This method combines the segmentation results of two steps for a more guaranteed performance. By comparison to other segmentation methods, the proposed methods show their effectiveness in different CBMIA systems.

- Object Tracking:
  An *Association Based Tracking Approach* to track multiple objects in a set of low frame rate videos is proposed in this work. The association is formulated as a "MAP problem" and solved by the Hungarian algorithm. The efficiency and effectiveness of the framework is verified on challenging micro-alike video datasets for insect behavioural experiments.

1.4   Overview of this Dissertation

Chapter 2 introduces an overview of existing CBMIA techniques and applications. This is important for understanding the classical and state-of-the-art algorithms in CBMIA development.

Chapter 3 describes a semi-automatic microscopic image classification system using an SSL framework, where image segmentation, shape features, classifier design and fusion are proposed and applied.

In Chapter 4, a full-automatic microscopic image classification system is introduced, which inserts SC features into a WSL framework, leading to a joint classifying and localising result.

Chapter 5 states an unsupervised learning system for microscopic image clustering work. In this system, image segmentation, shape feature extraction and unsupervised learning algorithms are developed and applied.

In Chapter 6, an interactive learning based object tracking technique is stated, where the object detection and frame-to-frame tracking algorithms are introduced continuously.

Experimental settings and results are presented in Chapter 7 by three practical CBMIA tasks to prove the effectiveness of the proposed methods. The first is EM classification, the second is stem cell analysis and clustering and the last is insect tracking. After the above tasks, several additional experiments are carried out to further verify the usefulness of these methods.

Finally, Chapter 8 closes the whole work by a conclusion. Furthermore, the future work is also put forward.

Chapter 2

Related Work

This chapter introduces the existing work that relates to CBMIA algorithms and applications. In Section 2.1, related algorithms are surveyed, including image segmentation, feature extraction, machine learning and object tracking techniques. Then, an overview of various practical CBMIA applications in the biological domain is presented in 2.2, where microorganism classification, cell clustering and insets tracking methods are discussed. Finally, a brief summary closes this chapter in Section 2.3.

2.1 Related Algorithms in CBMIA

2.1.1 Image Segmentation

Image segmentation is a digital image processing approach which partitions an image into multiple sets of pixels or regions, each possessing certain properties [SR09]. There exist different segmentation techniques, corresponding to different practical tasks and applications. These segmentation techniques are normally categorized into three types: The 'Thresholding or Clustering' method is useful for both colour and grey-level images in most cases [FM81; VS12]. The 'Edge Detecting' method is more suitable for grey-level images [FM81; SR09]. In contrast, the 'Region-based' method is more robust to colour images [FM81; VS12]. In the following part, existing pieces of work on these three image segmentation types are introduced.

Thresholding or Clustering Method

The thresholding method uses properties of intensity levels to distinguish different regions in a given image [FM81]. Traditionally, this is introduced to solve a grey-level image segmentation problem. In many practical tasks of image processing, the intensity levels (grey-levels) of pixels of an image's foreground (the interesting objects) are substantially different from those of its background (the remaining objects). Because of this difference, thresholding is used as an effective tool to separate the interesting objects from the background [SS04]. The segmentation result of the thresholding method is a binary image, where the foreground object and background are represented by two complementary states. Usually, the foreground is labeled by 1 (white) and the background is labeled by 0 (black), or conversely the foreground by 0 and the background by 1 [SS04]. For example, Otsu thresholding [Ots79] is an effective intensity-based segmentation method for grey-level images. This method minimises the weighted sum of variances of the foreground and background pixels for obtaining an optimal threshold. Especially when the amounts of pixels in foreground and background are similar, this method supports good edge detecting performances.

As a development of the traditional thresholding methods for grey-level images, many new colour image segmentation methods have been proposed in recent years. For instance, [CJW02] introduces a thresholding method for colour image segmentation, which analyses the occurrence of grey-levels of pixels in different colour channels simultaneously. This method first forms a histogram for each colour channel respectively, where the peaks of each histogram are used to establish a thresholding to obtain an initial segmented result. Lastly, these three initial results are united together to obtain the final segmentation result.

Furthermore, as an expansion of the thresholding method, the clustering method was developed. The clustering method typically uses more than two features to represent the characteristics of each class (cluster) of regions in an image [FM81; VS12]. For example, [SM00] proposes a clustering method, namely 'Normalised Cut', which extracts the global representation of an image. Normalised cut deals with the image segmentation problem as a graph partitioning task, where the total dissimilarity and similarity between different clusters is calculated to measure their consistencies.

Edge Detecting Method

In an image, an edge is a boundary between two different regions, where a sharp discontinuity of characteristics or computed properties occurs. Because different regions have respective characteristics, for example textures or colours, the edge is useful for segmenting different regions and localise objects [SR09]. Generally, to detect edges, three steps are needed: denoising, morphological operation and detection. Denoising: Most edge detectors work well with high quality images. However, because noise makes it difficult for edge detectors to distinguish edges of different regions, edge detectors often lose effectiveness when there is too much noise in the images. To reduce the influence of noise, various denoising techniques are usually applied before edge detection, e.g. the median filtering method. Morphological operation: To enhance the quality of digital images, morphological operations are used. For example, erosion and dilatation are often applied to smooth edges of objects to address an image sharpening problem. Detection: To detect edges, different algorithms are proposed. Generally, edge detection methods are grouped into two classes: 'Classical' methods and 'Soft Computing' approaches [LS10; SR09].

- Classical Methods:
  Sobel [Sob14], Prewitt [Pre70], Roberts [Rob63], Laplacian of a Gaussian (LoG) [Ney93], zero-crossing [Cla89], and Canny edge detectors [Can86] are a category of methods which analyses the local changing of images. For example, Roberts method monitors gradients of pixels to find edges, and judges a place where the maximal gradient occurs as the edge.

- Soft Computing Approaches:
  In contrast to the classical edge detecting methods, currently various soft computing methods are proposed to find edges in images. For example, fuzzy [Zad98], genetic algorithm [WZJ08], neural network [MC96], and wavelet [Niy+05] based approaches are often applied, which mainly use the methods of machine learning and signal processing.

Among all these edge detecting methods mentioned above, the Sobel edge detector is robust to the problem of noisy images because of its noise suppression functionality [Dre+07]. Hence, it is widely used in CBMIA tasks.

Region-based Method

Region-based segmentation methods divide an image into multiple regions by their spatial relationships of pixels. These kinds of methods group pixels into a region by the relationship between them and a standpoint [FM81; VS12]. For example, the watershed algorithm simulates topological features of geodesy and divides an image into different areas considering pixel values as altitudes [Cou+09]. Another famous instance is the region growing algorithm, which uses seed points for initialisation. Then, it determines whether its neighbouring pixels should be added to this region based on some given conditions, e.g. the work in [Fan+01].

2.1.2 Shape Features

To describe shapes of objects, different features (or descriptors) are developed. Generally, there are two kinds of shape features: 'Global' and 'Local' shape features. Global shape features represent an object by its characteristics of the entire object, e.g. its size and length. In contrast, local shape features describe the object by the partial characteristics of the object, e.g. the right-angles within this object. Although global and local shape features are opposites, they are able to be converted into one another. For example, the statistics of local features are usually used as global features, and global features can be applied to describe different parts of the object.

Global Shape Features

Global shape features include 'Contour-based' and 'Region-based' features [Mar04]. The former represent objects using characteristics of points on their shape boundaries. The latter apply properties of interior points to describe the objects [YKR08].

- Contour-based Features:
  There are many existing contour-based features, for example *Shape Signature* (SS) is an effective descriptor, where the set of centre distances of an object is used to describe its shape [ZL02]. Because the SS is suffering from the image rotation problem, the *Shape Signature Histogram* (SSH) method is proposed to solve this by quantising the signature into a rotationally invariant histogram [ZL04]. Besides SSH, *Fourier Descriptor* (FD) is an effective method to enhance the SS, which uses Fourier transformation to eliminate the rotation sensitiveness of SS [ZL03]. In [BMP02], *Shape Context Feature* (SCF) is proposed, which considers the correspondences between points on the contour of a shape. The *Edge Histogram Descriptor* (EHD) is effective to describe a shape using five directions of its edges [FG11]. In contrast to the basic EHD, the *Isomerous Edge Histogram Descriptor* (IEHD) models the distribution of edge lengths on the contour of an object [Li+13b]. Currently, the internal structure histogram (ISH) is developed and useful for distinguishing shapes by different internal structure angles [LSG15a].

- Region-based Features:
  There are also many region-based shape features, e.g. *Basic Geometrical Features* (BGFs), like area and length, are robust to measure a shape by its various geometrical values [Li+13a]. Based on the BGF, multiple *Higher-level Geometrical Features* (HGFs) are developed, just as different moment features. For instance, *Hu Moment* (Hu) is introduced in [Hu62], which can represent a shape by seven high order invariant moments. *Zernike Moment* (Zernike) is another robust invariant shape feature, which can yield a high quality set of orthogonal complex moments of a shape [TSS11].

Local Shape Features

In addition to the above-mentioned global features that represent overall characteristics of an object, it is valuable to use local features which describe characteristics of local regions in the object. Such local features capture the detailed shapes inside of an object. To this end, different methods are developed, for example the *Scale-invariant Feature Transform* (SIFT) feature represents the shape in a local region [Low04]. Because SIFT only describes local characteristics of an object, the *Pairwise SIFT* (PSIFT) feature is proposed to enhance the SIFT feature using spatial information and relations of the local characteristics [MS10].

2.1.3 Sparse Coding Features

Sparse coding (SC) represents a data vector by a sparse linear combination of bases, which is widely used in machine learning, pattern recognition and signal processing [Mai+10]. By using SC, raw pixel values of an image are converted into higher-level local features, which is useful for improving the effectiveness of feature extraction techniques [Rai+07; Lee+09]. For different applications and tasks, there are various sparse coding types.

Sparse Coding

Normally, SC is widely applied in audio, image and video retrieval tasks, e.g. [Plu+09] proposes a sparse representation approach for audio and music analysis. Because SC discovers bases that capture higher-level features in the data using only unlabeled input data, it provides a category of algorithms for exploring effective representations of inputting data. To use such SC, there is a computational problem to be solved where different methods are developed. For example, [Lee+07] presents efficient SC algorithms by iteratively solving an L1-regularized and an L2-constrained least squares problem.

Non-negative Sparse Coding

Though the normal SC can describe data effectively, in some situations it is restricted by non-negative conditions, e.g. in [LSG15c] SC needs to meet a non-negative upper bound condition for a fast searching algorithm. To this end, *Non-negative Sparse Coding* (NNSC) is proposed to meet this limitation. NNSC is an approach which decomposes multivariate data into non-negative sparse components. For example, an NNSC learning method is presented in [Hoy02], where an efficient multiplicative algorithm for finding the optimal values of the hidden components is proposed.

2.1.4   Supervised Learning

There are two types of supervised learning (classification) strategies, namely strongly supervised learning (SSL) [AL12] and weakly supervised learning (WSL) [SGU15]. The former uses strong annotations on the inputting data, where the exacted information of the data can be fully explored. In contrast, the latter only allows image annotation with the label of its category. WSL first extracts features from original images. Then, the extracted features are used to train a classifier by analysing the regional characteristics of the images. Finally, WSL can find the most interesting objects in each image and identify their classes.

Strongly Supervised Learning

SSL is the most used method in machine learning work. With a strong supervision, data supports more obvious characteristics than the original input. For example, [Nev+13] uses human interaction to pre-classify and segment images for a CBIR task, where a strongly supervising step incorporates prior information to the algorithm by the professional knowledge of humans. The strongly supervised data is used to build classifiers (classification algorithms) for identifying the data into corresponding categories. Furthermore, multiple fusion approaches are proposed to enhance the classification performance of single classifiers.

- Single Classifiers:
  There are many classification methods, for example, k-Nearest Neighbour (KNN) is a similarity-based non-parametric classifier. In KNN, the classifier consists of the k closest training examples in the feature space [CH67]. Another used method is the Naive Bayes algorithm, which builds a simple probabilistic classifier based on the Bayesian theory, where strong (naive) independence is assumed between the features [HY01]. In recent years, *Support Vector Machine* (SVM) is developed and applied widely [Bur98]. Compared to the SVM, similarity-based classifiers like KNN and probability-based classification methods like naive Bayes do not work well for high-dimensional features. Similarity-based classifiers fail to appropriately measure similarities in high-dimensional feature spaces, because of many irrelevant dimensions. Probability-based classifiers need a large number of image examples to appropriately estimate probabilistic distributions in high-dimensional feature spaces [GD05]. However, due to the practical tasks, it is usually difficult to collect a large and statistically relevant number of data for training. For these reasons, an SVM extracts a decision boundary between images of different data classes based on the margin maximisation principle. Due to this principle, the generalisation error of the

SVM is theoretically independent of the number of feature dimensions [Vap98]. Furthermore, a complex (non-linear) decision boundary can be extracted using a non-linear SVM to enhance the classification performance [CL11].

- Fusion Approaches:
  Besides the single classifiers mentioned above, different fusion methods are used in SSL tasks to improve the classification results. There are two essential types of feature fusion schemes, namely *Early Fusion* and *Late Fusion* [SWS05].

Early fusion combines all features into a single feature vector and performs classification based on it. The combination of features is realised by simply concatenating them [SWS05], or by matrix factorisation [JSD10; Gua+12]. However, high-dimensional feature vectors created by early fusion causes the 'curse of dimensionality' problem, where the performance of a classifier degrades when the dimensionality of a feature increases. In general, as the number of feature dimensions increases, the number of training images required to construct a well generalised classifier exponentially increases as well [JDM00]. To solve the curse of dimensionality problem in early fusion, some work carries out dimensionality reduction [Whi+12; XTX14]. However, because this reducing process always keeps main information of an image, it is easy to lose remaining important but detailed information, leading to an incomplete feature representation.

In contrast to early fusion, late fusion combines classification results obtained for different features. Specifically, late fusion exploits different features separately from each other, so it is more robust to the curse of dimensionality problem than early fusion. For instance, in recent work [YWT12; YRT14; XTX13], late fusion is used to enhance the classification and the clustering results in multi-view learning tasks, where features extracted from multiple camera views of a single object are effectively combined.

In addition to early and late fusions, middle fusion fuses different features during the process of training a classifier. For example, *Multiple Kernel Learning* (MKL) is a popular middle fusion approach [Xu+10]. In MKL, different kernels corresponding to different features are combined to increase the classification accuracy. Compared to single kernel classifiers like the SVM, MKL constructs a more powerful non-linear classifier which extracts a very complex classification boundary between positive and negative images. However, due to this excess complexity of the classifier, MKL often causes over-fitting. Especially in EM classification, the number of positive images is much smaller than negative images. This is because, except for the EM shown in positive images, all the other EMs need to be covered by negative images. In such a case, the classification boundary is excessively biased to the region of positive images (minority class) [HG09]. Furthermore, although the classification results based on a single kernel and MKL are both influenced by the over-fitting problem, existing experiments in [TCY09] prove that using only one kernel yields a more robust performance than using multiple kernels in the small dataset case. Especially in the binary classification tasks with less training data (around 25 training examples), applying only one kernel has a much higher performance than the one using MKL. When the number of training data is more than 750, MKL begins to show its advantage in feature fusion and has a better performance than one kernel. Based on the above discussion, late fusion is chosen for the CBMIA tasks in this work.

Weakly Supervised Learning

Although SSL can carry out very accurate information for classification work, it always costs large manpower and is time-consuming. To this end, WSL is developed and receives great contributions in object classification and localisation fields, where the majority of existing methods work on *Part-based Models* (PBMs) [FPZ03; PL11; SX11] and *Segmentation-based Models* (SBMs) [BU04; CL07; Gal+08; ADF10; Uij+13; Ngu+09].

- Part-based Models:
  The PBMs decompose an object into independent parts and extract features from each single part, then combine the features of each part to represent the whole object [Fel+10]. For example, [FPZ03] first uses an entropy-based feature detector to search an interesting region in a given image. Then, a probabilistic-based PBM is used to represent all parts of the object in this region. Finally, a Bayesian classifier is structured for classifying different images. In [PL11], each PBM first uses a lower-resolution root filter to represent the whole object, and uses a set of higher-resolution part filters to describe object parts; then, a latent SVM is trained to localise and recognise objects. [SX11] introduces a PBM framework based on the combination of an initial annotation model and a model drift detection method which obtains more effective localisation results.

- Segmentation-based Models:
  The SBMs exploit low-level image segmentation as a preprocessing phase before feature extraction, where objects are detected by the combination of different over-segmented regions in the images [Gal+08]. For instance, in [CL07] an SBM is used to classify different objects, where an image is first over-segmented, then, each region is only assigned to one object class, finally, a classifier is trained

using features extracted from these regions. [Uij+13] proposes an SBM based on the combination of an exhaustive search and image segmentation which can use the complementary of different segmented parts in an image to search the interesting objects. In [BU04], a class-based SBM is built by combining a top-down and a bottom-up segmentation processes, before extracting a set of informative regions as features and eventually building a classifier by a voting strategy considering the amount of present features. [ADF10] introduces an SBM which is established by considering the present possibility of an object with its surroundings together, where four complementary segmentation methods are integrated into a Bayesian framework to find the most promising region. [Ngu+09] proposes a weakly supervised *Region-based SVM* (RBSVM) classifier that is built to localise and classify an interesting object simultaneously, where a branch-and-bound searching method is effectively used [LBH08].

The majority of the existing WSL approaches use hand-crafted features. For example, [FPZ03] defines up to 30 partial shape features in the images. [PL11] uses a variation of Histogram-of-Gradient (HoG) features to represent objects. [SX11] applies *Bag-of-Visual-Words* (BoVW) features to represent each image. In [BU04], manually segmented shape features are defined to represent the presence or absence of an object. [CL07] uses BoVW features and partial shape features to describe the images. [Gal+08] applies BoVW features to represent each image. [ADF10] defines a set of HoG features to describe the images. In [Uij+13] and [Ngu+09], BoVW features are used to represent each image. Furthermore, the classification algorithms used in WSL are the same as those used in SSL, e.g. KNN and SVM.

2.1.5 Unsupervised Learning

Unsupervised learning (clustering) is able to recognise data in a full-automatic way without a classifier training process. There are two usual clustering methods: The first is the hierarchical clustering algorithm, which is a kind of nested clustering algorithm that builds a hierarchy of different classes (or clusters) [Sib73]. The second is the *k*-means clustering algorithm, which is a kind of partitioning clustering method and robust to deal with the large scale data case [Mac67].

Hierarchical Clustering

Agglomerative and divisive algorithms are two general types of hierarchical clustering methods.

- Agglomerative Algorithm:
  It is a 'bottom-up' approach where each data starts in its own cluster, and pairs

of clusters are merged (or combined) as one in the hierarchy. This process is iteratively done till each cluster is convergent. For example, [ZZW13] proposes an agglomerative algorithm based on a novel graph structural descriptor. This method assumes that when the descriptors of two clusters have substantial change during the process of merging, there exists a very close relationship between these two clusters, and they can be combined into one cluster.

- Divisive Algorithm:
  In contrast to the agglomerative algorithm, the divisive algorithm works in a 'top-down' way, where all data starts in one cluster, and is then splits into different individual clusters in the hierarchy. For instance, an effective divisive algorithm is introduced to solve a task of text classification in [DMK03], where a three-level hierarchy is built for a practical HTML document classification work.

k-means Clustering

k-means clustering partitions different data into k clusters, where each data belongs to the cluster with the closest mean value. To find the optimal clusters, an iterative refinement technique is the most commonly used, where the algorithm stops to find new clusters until all clusters have no essential changes. In contrast to the hierarchical clustering algorithm, k-means clustering has a faster computational speed when the number of data is large and the amount of clusters is small. Also, k-means methods can produce more convergent clusters than hierarchical clustering [SN14]. Due to these robust properties, k-means clustering methods are widely used in various tasks, for example:

- Classification:
  In [LW09], a new distributed k-means clustering method is introduced to classify tens of millions of phrases. In another task, k-means clustering is used to group large scale unlabeled text data into different categories [DM01].

- Feature Extraction:
  In recent years, k-means clustering [Csu+04] shows a brief and effective performance in the establishment of BoVW. Based on this work, [Blu+11] extends the feature learning approach to an object recognition task of RGB-D images.

2.1.6   Object Detection and Tracking

Object detection and tracking are very important in image series (or video) analysis. Object detection is used to verify existence of an object in the video. Object tracking is applied for following and localising the region of the object in the video.

Object Detection

Generally, there are three kinds of object detecting approaches, namely *Frame Difference*, *Optical Flow* and *Background Subtraction-based* methods. These methods are widely used in many tasks, for example:

- Frame Difference-based Method:
  The frame difference approach detects moving objects in a video, keeping the changed regions (or pixels) and eliminating the unchanged regions. Because this method is simple to implement and has a great detecting speed, it is widely used. For example, in [Zha+07] an improved frame difference algorithm is introduced to detect moving objects. It first detects the edges of each two continuous frames by an edge detector and obtains the difference between these edge images. Then, it divides the obtained edge images into small blocks and compares them in order to decide whether or not they are moving areas. Finally, it connects the blocks to obtaining the smallest bounding box in which the object is included.

- Optical Flow-based Method:
  Optical flow is a descriptor of a moving object in a given scene, which describes the relative motion between an observer (e.g. a camera) and the scene [WS85]. For instance, a moving object segmentation method is proposed based on the optical flow method [Kla+09]. This approach first provides a motion metric that corresponds to the likelihood of the moving object. Then, using this metric, the object is segmented by a globally optimal graph-cut algorithm.

- Background Subtraction-based Method:
  The background subtraction algorithm defines the background of a scene as an immobile part in a video, which is used to detect moving objects in the video. For example, in [SJK09] the background subtraction algorithm is extended to capture objects from freely moving cameras.

Object Tracking

Normally, three types of object tracking methods are used: *Point*, *Kernel*, and *Silhouette-based* algorithms.

- Point-based Method:
  In point-based tracking, key-points (or interest points) are first detected. Then, based on the characteristics of these key-points, the corresponding objects are tracked. For example, a point-based tracking method is combined into a multiple hypothesis tracking framework in [TS05]. In this work, an object is first

described by features of its key-points. Then, the key-points are detected using edge and corner detectors. Lastly, these points are tracked frame-by-frame in the video to show the tracking result of the object.

- Kernel-based Method:
  Kernel-based tracking approaches iteratively use a localisation procedure to find an object with maximal similarity to the tracked object in the last frame, where the object is represented by the kernel algorithm in each frame. For instance, [CRM03] proposes a visual tracking method, where the target object is first represented using a spatial masking with an isotropic kernel. Then, the target localisation is solved by a spatially-smooth similarity function of the masking. Finally, the effectiveness of this method is proven in two object tracking tasks.

- Silhouette-based Method:
  Silhouette-based (or contour-based) tracking methods use the previous frames to generate an object model, and use this model to find the object region in the current frame. For example, [Ros+05] proposes a silhouette-based human motion tracking system, where a free-form surface model is automatically generated using a set of input images.

Based on the typical tracking methods mentioned above, *Multiple Object Tracking* (MOT) methods are developed. The MOT approaches could be classified into two categories [LZK14]: Association based and *Category Free* tracking approaches.

- Association based Method:
  These types of methods usually first localise objects in each frame and then link these object hypotheses into trajectories without any initialization labeling, which can track varying numbers of objects. The success of most existing association based tracking algorithms comes from the discriminative appearance model (using the colour or texture features) [QS12; HWN08], or the constant velocity motion of targets [BWG07; QS12; Per+06; HWN08].

- Category Free Method:
  Category free tracking methods are referred as online object tracking [WLY13], requiring the initialization of a fixed number of objects in the first frame (in the form of bounding boxes or other shape configurations), then localize these fixed number of objects in the subsequent frames.

A more detailed summary of the related tracking approaches to this dissertation is given in Table 2.1, including the tracking framework, appearance model, type or number of target(s), and the assumptions of these works.

Table 2.1: Object tracking related work and main characteristics.

	Tracking framework	Appearance model	Type/number of target(s)	Remarks
[BWG07]	Hungarian	Foreground response	Multiple generic objects	Assume coherent motion
[QS12]	Hungarian	Colour histogram	Multiple human pedestrians	Assume coherent motion
[Per+06]	Hungarian	Foreground response	Multiple cars	Assume coherent motion
[HWN08]	Hungarian	Colour histogram	Multiple human pedestrians	Assume coherent motion
[VCS08]	Particle filter	Geometrical features	Single bee	Incorporate specific behavior model
[LR07]	Particle filter	Optical flow	Single bee	Assume coherent motion
[Pis+10]	Particle filter	Foreground response	Multiple mice and larvae	Assume coherent motion
[BKV01]	Simple data association technique	Foreground response	Multiple ants	Does not tackle occlusion and merges
[Yin04]	Particle filter	Foreground response	Multiple ants	Assume coherent motion
[Ris+13]	Not specified	Specific warm-like insect features	Multiple Drosophila larvae	Cannot resolve collisions involving more than two animals
[Bra+09]	Hungarian	Area of connected components	Multiple Drosophila adults	Assume coherent motion
[Fia+14]	Graph based framework	Combined features that capture local spatiotemporal structure	Multiple Drosophila larvae	Training samples of encounters of two larvae required
[Muj+12]	Antennae identified by two largest clusters	None	Two bee antennae	Does not tackle MOT problems including merge, occlusion, etc
[VSC08]	Probabilistic framework	Splines	mouse whiskers	Does not tackle MOT problems including merge, occlusion, etc
[PE+14]	Probabilistic framework		Constant number of animals	Assume restrictive criterion

2.2 Biological Applications Using CBMIA

In recent years, CBMIA is widely applied in biological domains, for example environmental biology, bio-medicine and bio-agriculture. In the following part, a brief overview of the biological applications using CBMIA techniques is given from three aspects: Microorganism classification, cell clustering and insect tracking.

2.2.1 Microorganism Classification

Traditionally, microorganism classification is done by chemical (e.g. chemical component analysis), physical (e.g. spectrum analysis), molecular biological (e.g. DNA and RNA analysis), and morphological (e.g. manual observation under a microscope) methods. For example, in [Mac+96] a chemical method is used to identify the classes of different phytoplankton, where the composition of pigments of each class is tested as the discriminative property. [Kir+00] tests species of microorganisms using a physical approach. In this work, the similarity of fourth derivative spectrum between a test sample and a standard sample is used to confirm the class of this test sample. In [GSL91; Ber+95], molecular biological techniques are introduced to accurately identify different microorganisms by their DNA or RNA principles. The morphological method is the most direct and brief approach, where a microorganism is observed under a microscope and recognised manually based on its shape [PGG14]. These traditional methods have different working mechanisms and results, where the main advantages and disadvantages of them are compared in Table 2.2.

Table 2.2: A comparison of traditional methods for microscopic image analysis.

Methods	Advantages	Disadvantages
Chemical method	High accuracy	Secondary pollution of chemical reagent
Physical method	High accuracy	Expensive equipment
Molecular biological method	High accuracy	Secondary pollution of chemical reagent, expensive equipment, long testing time
Morphological method	Short testing time	Long training time for skillful operators

From Table 2.2, people can find that these traditional microorganism classification methods suffer from three respects: Secondary pollution, expensive equipment and long duration time. In order to solve problems of the above mentioned traditional methods, CBMIA approaches are applied to support a cleaner, cheaper, and more rapid way for microorganism classification tasks. In this section, CBMIA based microorganism classification work is reviewed by the applying purposes, where agricultural, food, industrial, medical, water-borne and environmental microorganisms are referred to.

Agricultural Microorganism Classification

Agricultural Microorganisms (AMs) are the investigative objects of agricultural microbiology, which are associated to animal and plant diseases, soil fertility and soil structure [RB04]. Beneficial AMs can help farmers to increase the agricultural yields, for example *Rhizobia* can obtain nitrogen from soil and fix it into root nodules of legumes. Harmful AMs can bring diseases to crops and domestic animals, for instance the tobacco mosaic virus can infect leaves of tobacco and reduce harvest seriously. To this end, AM classification is important to distinguish different AMs in order to guarantee and increase the agricultural production.

Furthermore, the CBMIA approaches for AM classification is introduced because of its effectiveness. For example, in [Liu+01] an image analysis system is established to classify AMs into different categories. In this system, image segmentation, shape feature extraction and a shape classifier are used. Finally, 11 classes of microorganisms are classified, leading to a classification accuracy of 97.0%, where around 1400 microorganisms are used for training the classifier, and around 4200 microorganism are used for testing. In [Daz10], an upgrade of the aforementioned image analysis system is released, where grey-level and colour information is included in the segmentation process to increase the overall segmentation accuracy. Lastly, the system obtains a better performance, where the classification accuracy is higher than 99.0%.

Food Microorganism Classification

Food microbiology studies the habits and usages of *Food Microorganisms* (FMs) that appear in the producing, transporting, and storing processes of food [JLG05]. "Good" FMs can help humans to produce nutritious and delicious foods, for example *Lactobacillus* is employed in dairy production for making yogurt and cheese. In contrast, "bad" FMs can result in food deterioration, for instance *Clostridium* can cause fish to decay and become smelly. Hence, the classification of different FMs plays a very important role in the daily life of humans.

To increase the effectiveness of FM classification work, CBMIA techniques are applied. For instance, in [KM08] shape and colour features are used to classify five categories of FMs, where the similarity of each feature is used as the measurement of classification. In this work, 34 images of five FM classes are used for experiments, and around 90% mean accuracy is obtained. The framework of this work is also used in [KM09], where textural features are applied instead of shape and colour features, and finally a mean classification accuracy around 94% is achieved. Based on these two previous works, [KM10] integrates shape, colour and textural features, and uses a probabilistic neural network classier for classification. Lastly, the mean classification accuracy of these FMs is improved to 100%.

Industrial Microorganism Classification

The microorganisms work in industrial fields are called *Industrial Microorganisms* (IMs), which are the investigative objects of industrial microbiology [Oka07]. On the one hand, helpful IMs are able to yield industrial products for human beings, e.g. *Aspergillus niger* can make citric acid very effective. On the other hand, undesirable IMs always disturb industrial processes, for example sulfate-reducing bacteria can corrode ironwork, leading to the damage of mechanical equipment. Because of the significance of IMs mentioned above, IM classification is needed to distinguish different categories of IMs for guaranteeing industrial activities.

Because IM classification is very important, various techniques are applied to deal with it, and recently, CBMIA has been introduced to solve this problem. For instance, in [FH89] image segmentation, shape feature extraction and a similarity-based matching algorithm (namely orderly relationship) are used in an IM classification task. In the experiment, six classes with three to five examples for each class of IMs are tested, where a mean classification accuracy of 75% is finally obtained. [CT92] proposes a method for an IM classification work, where skeletonisation and morphological operations are used for segmenting IMs, shape characteristics are used as features, and three geometrical ratios are used as discriminative threshold values to distinguish different IM categories. Finally, 49 IM examples are tested in the experiment, and a classification accuracy between 30% to 60% is obtained using three thresholds, respectively. Another example is [DJJ94], which classifies two categories of IMs using image segmentation, shape features and a geometrical value based linear classifier. In the experiments, 323 IMs of two classes are tested, leading to a classification accuracy of 100%.

Medical Microorganism Classification

Medical microbiology is an interdiscipline of medicine and biology, concerned with the prevention, diagnosis and treatment of infectious diseases to improve the health condition of a human [GB12]. In medical microbiology, *Medical Microorganisms* (MMs) are the research emphasis, which can both lead to and cure diseases. Pathogenic MMs make people sick, e.g. the influenza virus results in the common flu. In contrast, curative MMs can help patients to recover soon, for example *Penicillium chrysogenum* can produce penicillin helping a patient come back to full health. Hence, to obtain more knowledge of MMs to improve a human's health, MM classification plays a significant role in medical microbiology.

Because MM classification is very important, different methods are used, including the rapidly developing CBMIA approaches. For example, [RKG92] introduces an MM classification method using image segmentation, shape features and a linear classifier. This work classifies two categories of MMs in a qualitative way, so it does not evaluate the classification result (e.g. accuracy) with quantised measures. In [VCL98], edge detection, shape feature extraction and different classifiers are used to distinguish one class of MM from other objects. In this work, 1000 images are used for classifier training and 147 images are used for testing. Finally, the best classification result is obtained by a multi-layer neural network classifier, where an accuracy of 97.9% is achieved. [AB+00] proposes a system to specially identify a class of MM, where three colour channels, shape features and colour channel based threshold are used. Finally, the qualitative experimental result shows that the green channel obtains the highest classification performance, because it can support a high discrimination threshold to identify the MM, but there is no quantised evaluation results in this work. The work in [Tru+01] applies image segmentation, shape features and a minimum-distance classifier for a two-class MM classification task. In the experiment, ten images of each class are used for training the classifier, and 758 MMs are used for testing. Finally, an error ratio of 7.30% is obtained as the evaluation of the classification result. In [FV+02], a two-class MM classification work is done, where image segmentation, various features (shape, colour and texture), and a Fisher linear discriminant classifier are used. To evaluate this work, 130 images are used for classifier training, and finally a specificity of 82% is produced as the measurement of the classification result. Furthermore, an extension of this work is introduced in [FCA03], where a classification tree is used as the classifier and makes the specificity increase to around 92%. [Rod+04] tests an MM classification method on five classes of MMs, where amassing of elaborate knowledge is used in the final decision process, so this method is different from typical pattern recognition based CBMIA approaches, but more similar to an expert system. In the experiment, all single MMs are correctly classified, but collective ones are hard to be classified into the right categories. In [WSP05], an MM classification system is built up using a dither filter, shape and textural features, and an artificial neural network classifier. To test the effectiveness of this system, around 4000 images of two classes of MMs are used for training the classifier, and around 800 images are used for testing. Finally, this system receives a classification accuracy of 95.7%. [FCD06] introduces an MM classification approach, where image segmentation, shape and colour features, and a minimun error Bayesian classifier are used. In the experiment, around 900 images of MMs are used to train the classifier, and around 650 images are used in test. Finally, the classification result is evaluated by specificity and sensitivity, where they reach 98% and 94%, respectively. In [DLW06], an MM classification method is developed by image segmentation, shape and colour features, and geometrical thresholds, where a qualitative experimental result is obtained based on nine classes of MMs. A 3-D classification system is creatively proposed in [Jav+06], which uses multiple sensors and features to enhance the final classification performance. In this work, image segmentation is first applied, then shape and Gabor-based wavelets features are used to describe the MMs, finally, graph matching and statistical cost functions are used to classify different MM categories. In the experimental part, this system is proved by its effective classification result on three classes of MMs (each class 200 testing MM images), where T-test values are used in evaluation and the overall performance is around 55% (true class). In [Li+07], a global and local feature fusion technique is used to classify three categories of MMs, where a shape feature is used as the global feature and a spatial feature is used as the local feature. Furthermore, a matching approach is applied to classify the MMs. In the experiment, 800 images are used for testing, and a classification accuracy around 88% is finally achieved. An MM classification method is proposed in [MAA09], where image segmentation, shape and colour features, and a threshold are used. In its experiment, 300 images of one MM class are qualitatively tested. In [RSM11], image segmentation, shape features and a neural network classifier are used in a two-class MM classification task. In this work, 75 MM images are used for training the classifier, 25 images are used for testing, and finally a mean square error of 0.0368% is obtained as the overall evaluation of the classification result.

Water-borne Microorganism Classification

Water-borne Microorganisms (WMs) are the studying objects of water microbiology (or aquatic microbiology), which are found in different waters [MH03]. Useful WMs bring various benefits to human production, for example *Rotifera* is a good fodder in the fishing industry. In contrast, profitless WMs always impact producing activities and spread diseases, for instance *Shigella* causes bacillary dysentery through dirty water. So, WMs play a very important role in people's daily life, and the classification of WMs becomes the fundamental work of WM research.

Because WM classification is very necessary, different methods are applied, where CBMIA approaches show effective performance among all existing methods. For example, a WM classification approach is proposed in [USU78], where image denoising, shape feature extraction and geometrical thresholds are used for a five-class classification task. In this work, 15 WMs of each class are used for calculating the thresholds, five WMs of each class are used for testing, finally an accuracy of 100% is achieved. [TN84] introduces a system to classify two categories of WMs, where shape features and geometrical thresholds are used. In the experiment, 800 WMs of two classes are used for testing the classification performance of the system, and finally a mean mis-classification rate which is lower than 1.1% is calculated for evaluation. In [Jef+84], eight classes of WMs are classified with image segmentation, shape features and descriptive geometrical ratios. To evaluate the effectiveness of this work, 265 WM images are used for training and 50 images are used for testing. Lastly, a mean classification accuracy of 89% is obtained. A WM classification approach is developed in [Ish+87], which uses shape features and geometrical thresholds to classify WMs. In this approach, 17 WMs of a WM class are used for testing, where a classification accuracy around 95% is obtained. In [Bal+92], a method is proposed to classify eight classes of WMs, where shape features and a neural network classifier are used. In the experiment, 2000 images are used for classifier training, and finally a general classification accuracy that is higher than 90% is obtained. [Cul+94] develops a system to classify five categories of WMs. In this work, raw pixel values of a WM image are first converted by Fourier transformation and used as the feature of this image. Then, a classifier is built up using a neural network method. Finally, experiments prove the usefulness of this system, where 201 images are used for classifier training and 299 images are used for testing, and an error rate of 11% is achieved. A two-class WM classification approach is introduced in [TW95], where image segmentation, shape and textural features, and discriminative values are applied). In the experiment, 91.1% classification accuracy is obtained on 34 test WM images. In [Tru+96], image segmentation, shape feature extraction and discriminant functions are used in a WM classification task, where seven categories of WMs are classified. Finally, around 230 testing images are used in the experiment, leading to a classification accuracy of 80.3%. [Cul+96] compares the classification performance of four different classifiers in a WM classification task, including multi-layer perceptron network, radial basis function network, *k*-nearest neighbour and quadratic discriminant analysis algorithms. In this research, basic pixel based features, like pixel values converted by fast Fourier transform, are extracted to represent the WM images. Finally, in the experiment 23 classes (in total more than 5000 WM images) are tested, and the radial basis function network shows the best classification performance of 83% mean classification accuracy. In [TS96; Tan+98], a WM classification approach is used to distinguish six classes of WMs, where image segmentation, shape and textural features, and a novel parallel-training learning vector quantization network classifier are applied. In this work, 1000 WM images are used for classifier training and another 1000 are used for testing, and finally a classification accuracy of 95% is achieved. A WM classification method is introduced in [Bla+98], where object edge detection, shape feature extraction and an artificial neural network classifier are used. In this work, 60 images of three WM classes are used for training the classifier, and finally a classification accuracy higher than 90% is obtained. [KLW99] develops a system to classify four classes of WMs, where image segmentation, energy based features and energy thresholds are used to distinguish around 1900 WM images. Finally, a mean classification accuracy higher than 90% is achieved. In [Rod+01] a WM classification work is done where four classes of WMs are classified using image segmentation, shape features, and geometrical thresholds. To evaluate this method, seven WM sample sets are tested. A six-class WM classification work is introduced in [Cul+03], which uses shape features and a machine learning system to identify different WMs. In the experiment, 128 WM images are used for training the classifier, and 182 images are used for testing. Finally, an overall classification accuracy of 82% is obtained. In [EGH03], a WM classification approach is used to classify four classes of WMs. In this work, image segmentation, shape and textural features, and a neural network classifier are used for classifying WMs. Lastly, around 1700 images are used to test the classification performance of this approach, and a mean accuracy of 78.66% is achieved. A 3-D WM classification system is developed to classify two categories of WMs, using image segmentation, shape features and graph matching methods [Jav+05]. In the experiment, qualitative classification results are given to show the performance of this system. As an extension of this work, [YMJ06] does quantized evaluation for the system, where mean-squared distance (MSD) and mean-absolute distance (MAD) are used to measure the classification effectiveness. In this extended work, three classes of WMs are tested, where 100 positive and 100 negative images of each class are used for testing. Finally, mean MSD around 1.5% and mean MAD around 10% are achieved. An optic based WM classification system is developed in [Chi+11], where shape features and geometrical thresholds are used to classify two categories of WMs. Finally, 300 WM images of each class are used for calculating the geometrical characteristics in a quantitative experiment.

Environmental Microorganism Classification

In environmental microbiology, environmental microorganism (EM) is the general term that describes microscopic organisms living in natural environments (e.g. forests) and artificial environments (e.g. farmlands), which are useful for protecting and cleaning the environment [PGG14]. There are different usages of EMs, for example *Actinophrys* can digest the organic waste in sludge and increase the quality of fresh water. *Rotifera* can decompose rubbish in water and reduce the level of eutrophication. Hence, to achieve the environmental treatments, the classification of EMs is necessary and important.

Because the EM classification is very useful, effectively classifying EMs becomes an urgent need in environmental science and engineering. To this end, CBMIA approaches are developed and used to classify EMs. For example, [Li+13b] compares different shape features in an EM classification task, where a *Multi-class Support Vector Machine* (MSVM) classifier is applied. In the experiment, ten classes of EMs are used to test the effectiveness of different shape features. First, ten images from each EM class are used for training the classifier, and another ten images from each class are used for testing. Then, the classification performances of four shape features are compared, where classification accuracy is calculated as the measurement. Finally, geometrical features provide the best classification result (89.7% accuracy). Based on the work above, a semi-automatic EM classification system is established in [Li+13a], where a Sobel edge detector based semi-automatic image segmentation method is developed. Lastly, a classification accuracy of 66% is obtained using the semi-automatically segmented images. With the same experimental setting, another shape feature is tested in this EM classification system, leading to an accuracy of 79.5% [Yan+14]. In [LSG15a] an EM classification approach is proposed, using image segmentation, shape feature extraction, SVMs and late fusion techniques. In this work, ten categories of EMs are used to test the classification performance of the system, where ten images from each class are used for SVM classifier training and another ten images from each class are used for testing, and the mean of *Average Precision* (AP) is used as the overall evaluation of this EM classification task. Finally, 94.75% mean AP is achieved. As an extension of the above mentioned work, [Zou+15] introduces an EM image retrieval system, where image segmentation, ISH shape descriptor and similarity matching methods are applied. In the experiment, 420 EM images (21 classes, each class 20 images) are tested, and finally a 33.9% mean AP is obtained. A WSL EM classification method is proposed in [LSG15c], using NNSC and BoVW features, RBSVMs and late fusion approaches. This work can localise and classify EMs jointly, leading to a full-automatic EM classification approach. In the experiment 15 classes of EMs are used, where ten images of each class are used for classifier training and another ten images of each class are used for testing the performance of the system. Finally, around 60% mean AP is achieved. In [LSG15b], an EM classification work is introduced, using image segmentation, global and local shape features, SVM classifiers and late fusion approaches. In the experiment, 15 categories of EM are tested, where ten image of each class are used for training the classifiers and another ten images of each class are used for the classification test. Lastly, 83.78% mean AP is obtained.

2.2.2   Cell Clustering

In biology, cells are the basic units of organisms which can structure living bodies and implement biological functions [Sch93]. On the one hand, cells can structure the bodies of different organisms, for example *Euglena* is a unicellular organism. On the other hand, cells can carry out different biological functions, for instance the cells of plant leaves can process photosynthesis to fix carbon. Thus, cells play a fundamental role in biological activities, which are useful in medical treatment, environmental engineering, and so on. To this end, effective CBMIA approaches are introduced to aid biological researchers to get to know more about cells. Among all the cell CBMIA work, cell clustering is a main task which classifies cells into different groups using an unsupervised learning approach. Cell clustering can help biological scientists to explore more unknown information of the cells, for example cancer cell clustering can conclude the differences between cancer and normal cells. Furthermore, a brief overview of CBMIA based cell clustering work is given as follows.

In [Bei+06], image segmentation, shape features and a simple hierarchical agglomerative clustering method are integrated into a framework of Minkowski valuations for a neuronal cell analysis task. In this work, the neuronal cells are classified into three types based on their shapes and the hierarchical clustering approach. A clustering work of breast epithelial cells is proposed in [Lon+07], where shape features are used to describe the cells and five clustering methods (e.g. *k*-means and hierarchical clustering) are combined to enhance the clustering performance. In the experiment, eight clusters of cells are finally obtained. [Cao+08] introduces an approach to analyse the transcriptional and functional profiling of human embryonic stem cell-derived cardiomyocytes. In this work, energy based features and *k*-means clustering are used to aid the analysing process. In [Low+08], a hierarchical clustering approach is used to indicate the types of stem cells from dermal fibroblasts. A lung branching morphogenesis analysis work is proposed in [Rod+14], where a clustering approach is applied to classify the inner lung epithelium into different groups. [Sar+14] uses a *k*-means clustering approach in an analysis of leukocytes. In this work, the *k*-means method is applied in image segmentation and classification of two sub-tasks. In [Hig+15], a clustering approach is used to distinguish interesting cells from other objects and the background, where shape features are used as the descriptors to represent the cells.

2.2.3 Insect Tracking

Insects are the most prosperous groups of animals on the earth, containing millions of species and playing an important role in human life [GE05]. On the one hand, beneficial insects can help people to put out products in agriculture and industry, for example honey bees can carry pollen from flower to flower to accelerate the cross-pollination (the fertilization process of plants). On the other hand, harmful insects can spread diseases and destroy farming gains, for instance flies and mosquitoes are the most common spreaders of diseases. Hence, in order to obtain the best input from different insects, people need to analyse the behaviour and characteristics of insects. To this end, CBMIA is introduced to aid biological researchers to learn more about insects. Especially because insects are always moving, tracking them is very important in the analysing tasks for capturing the dynamic details. There are two types of insects tracking tasks: One is to track the entire insects, and another is to track the body parts of insects.

- Tracking Entire Insects:
  There are a few published studies that address problems about insect tracking. For example, [LR07] introduces an algorithm for tracking freely moving bees in the famous "bee dance" process. In this work, an optical flow-based method is first used to detect the bees in each video frame, and then the detected bees are tracked by linking the frames together. In [VCS08], another algorithm is proposed to a similar task, where the shapes and behaviours of dancing bees are represented by an encoding approach for tracking the bees. An ant tracking work is presented in [BKV01], where a colour-based tracking method is combined with a movement-based tracking method for tracking hundreds of fast moving ants. [Yin04] develops an automatic ant tracking method, which tracks moving ants from video-clips to obtain their moving trajectories.

- Tracking Body Parts of Insects:
  Besides the tracking work of the entire insects, tracking the freely moving body parts of insects is another hot research point. For example, in [Muj+12], a point-based tracking method is used in a tracking task of bees' antennae. [She+13] introduces an automatic method to track different body parts of a bee. This work incorporates prior information about the kinematics and shapes of body parts for a multiple object tracking aim. In [She+14], a multiple tracking work is proposed to track bees' body parts, where an association based method is

applied in an interactive framework. Based on the work above, a practical system is established in [She+15a], which supports a user-friendly application interface. [She+15b] introduces an association based tracking approach to track multiple insect body parts in a set of low frame rate videos. A more detailed review can be found in [PE+14].

2.3   Summary

This chapter introduces the related work of CBMIA approaches. In the first part, state-of-the-art CBMIA techniques are briefly discussed, including image segmentation, feature extraction, machine learning, object detection and tracking (see Section 2.1). In the second part, an overview of CBMIA applications is given, especially focusing on microorganism classification, cell clustering and insect tracking tasks (see Section 2.2).

Chapter 3

Semi-automatic Microscopic Image Classification Using Strongly Supervised Learning

In this chapter, a microscopic image classification system using a strongly supervised learning (SSL) framework is proposed. In Section 3.1, an overview of this system is given first. Then, in this system multiple CBMIA techniques are used, including image segmentation in Section 3.2, global shape feature extraction in Section 3.3, local shape feature extraction in Section 3.4, and classifier design and fusion in Section 3.5. Finally, Section 3.6 closes this chapter with a brief conclusion. Moreover, an environmental microorganism (EM) classification system is structured in this chapter as an illustration.

3.1   Overview of the System

Figure 3.1 illustrates an overview of the proposed microscopic image classification system using SSL framework on EM images. It simulates the morphological method where environmental scientists detect classes of EMs based on their shapes. Given EM images, the system first conducts semi-automatic image segmentation to obtain regions of EMs. Then, the region of each EM is represented by features which characterise this region. Based on such representations, EM classification is considered as a binary classification problem, where images containing a certain EM are distinguished from the others. Two types of training images are used, *positive images* where the EM is present, and *negative images* where it is absent (and other EMs are present). By contrasting these two types of training images, a *Radial Basis Function Kernel SVM* (RBFSVM) classifier is built to discriminate between test images where

the EM appears and the other test images. This system is developed by addressing
the following three problems:

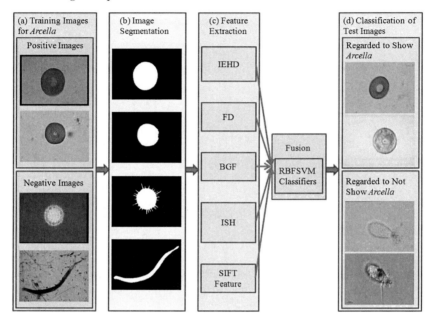

Figure 3.1: An overview of the EM classification system using SSL framework.

- Image Segmentation:
  To remove impurities in microscopic images, image segmentation is conducted.
  This can obtain the high quality shape information about the region of an EM.
  To this end, the semi-automatic segmentation is proposed in Section 3.2, which
  refines an EM region that is initially roughly specified by a user.

- Feature Extraction:
  To characterise EMs irrespective of colours and rotations, BGFs which represent geometrical properties of their regions are extracted, such as sizes, areas
  and shapes in Section 3.3.2. Furthermore, IEHD (see Section 3.3.1), FD (see
  Section 3.3.3) and SIFT feature (see Section 3.4) are also selected for this task.
  In particular, the structure of an EM is an effective criterion for EM classification. For example, *Arcella* (depicted by two positive images in Figure 3.1 (a))

and *Actinophrys* (depicted by the upper negative image in Figure 3.1 (a)) have similar sizes and shapes, but the latter is characterised by many pseudopodia. To capture such a structure, the ISH is proposed in Section 3.3.4.

- Fusion:
  Using only one single feature is not sufficient for accurate EM classification. For example, while ISH is useful for identifying *Actinophrys* with the distinctive structure, *Siprostomum* shown as the bottom negative image in Figure 3.1 (a) is accurately categorised based on its characteristic length-width ratio. Hence, how to fuse different features to achieve accurate EM classification is investigated. Regarding this, while fusion enhances the discriminative power of a classifier, it increases the risk that the classifier over-fits to training images. In particular, it is difficult to prepare a large number of training images for each EM class. Thus, fusion may cause a classifier which emphasises features very specific to a small number of training images, but useless to test images. To alleviate over-fitting, a late fusion approach (see Section 3.5.3) is used, which combines outcomes of classifiers based on different features by weighting them using the correspondence accuracy. In other words, over-fitting is avoided by building each classifier based only on a single feature, and combine these "not over-fit" classifiers.

3.2 Image Segmentation

This section first introduces the principle of Sobel edge detection and morphological operations in Section 3.2.1. Then, based on Sobel edge detection, a semi-automatic image segmentation method is developed in Section 3.2.2.

3.2.1 Sobel Edge Detection and Morphological Operations

Foundation of Mathematics for Sobel Edge Detection

The following part explains the gradient theory [GW08], which is the theoretical basis of Sobel edge detection. Assume $f(x, y)$ represents the intensity value at the position (x, y) in an original grey-level image. The gradient is defined as

$$\nabla f = (G_x, G_y)^{\mathrm{T}} = \left(\frac{\partial f}{\partial x}, \frac{\partial f}{\partial y}\right)^{\mathrm{T}}, \tag{3.1}$$

where the magnitude of the gradient vector is given by

$$\nabla f = \text{mag}(\nabla f) = (G_x^2 + G_y^2)^{\frac{1}{2}} = \left[\left(\frac{\partial f}{\partial x} \right)^2 + \left(\frac{\partial f}{\partial y} \right)^2 \right]^{\frac{1}{2}} . \tag{3.2}$$

In order to simplify the computation, the magnitude is approximated with

$$\nabla f \approx |G_x| + |G_y| . \tag{3.3}$$

This approximation still behaves as derivative, because it becomes zero in areas of constant intensity and it is proportional to the degree of intensity change in areas of varying pixel values.

A fundamental property of the gradient vector is that it represents the direction of the maximum change of f at the position (x, y). The orientation of this maximum change can be expressed by the following angle

$$\alpha(x, y) = \arctan \left(\frac{G_y}{G_x} \right) . \tag{3.4}$$

Based on the gradient theory mentioned above, the place where a sharp changing of pixel values occurs can be considered as an edge in the image. A visible example is shown in Figure 3.2

Technological Process of Sobel Edge Detection

Sobel edge detection is defined based on the aforementioned gradient theory, where two mask matrixes (Sobel edge detectors) are especially used to find the locations of edges in an image [Sob14]. The first mask matrix is related to G_x on the x-direction, and the second mask matrix is related to G_y on the y-direction. These two matrixes are shown in Figure 3.3.

To use these two masks to detect edges, the pixel values in an image patch is first represented as shown in Figure 3.4.

Then, the gradient ∇f can be approximated as follows [1]

$$\begin{aligned}
\nabla f &= |G_x| + |G_y| \\
&= |f(1,3) + f(2,3) * 2 + f(3,3) - f(1,1) - f(2,1) * 2 - f(3,1)| \\
&\quad + |f(3,1) + f(3,2) * 2 + f(3,3) - f(1,1) - f(1,2) * 2 - f(1,3)| .
\end{aligned} \tag{3.5}$$

Finally, this obtained gradient ∇f is compared to a given threshold, where if ∇f is larger than the threshold, this pixel is judged as an edge and set to 1 (white).

---

[1] In Eq. 3.5, the algorithm of convolution is discussed in [Rad72].

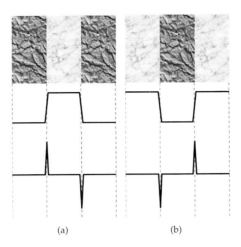

(a)   (b)

Figure 3.2: An example of gradient theory for edge detection. (a) A brighter edge between two darker parts. (b) A darker edge between two brighter parts. The top row shows the original images. The middle row shows the overall changing trends of the pixel values ignoring noise, where brighter parts have higher pixel values than darker parts. The bottom row shows the locations where the edges occur.

-1	0	1
-2	0	2
-1	0	1

-1	-2	-1
0	0	0
1	2	1

(a)   (b)

Figure 3.3: Mask matrixes of Sobel edge detectors. (a) The mask matrix for x-direction. (b) The mask matrix for y-direction.

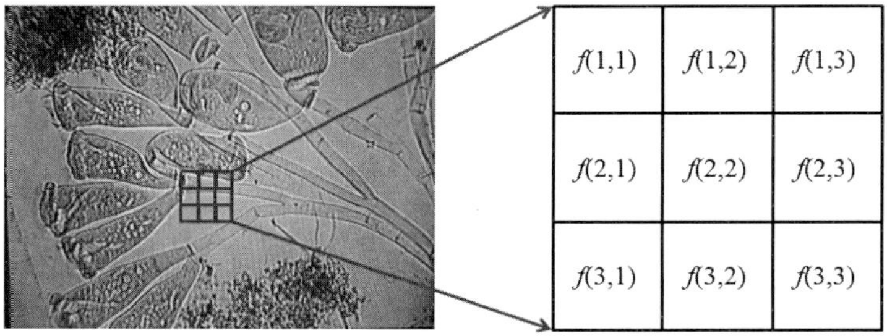

$f(1,1)$	$f(1,2)$	$f(1,3)$
$f(2,1)$	$f(2,2)$	$f(2,3)$
$f(3,1)$	$f(3,2)$	$f(3,3)$

Figure 3.4: An example for pixel values in an image patch. The left image shows an original image of *Vorticella*, where nine pixels in red boxes constitute an image patch. The left matrix shows the pixel values of the left image patch.

Conversely, if ∇f is smaller than the threshold, this pixel is judged as a part of the background and set to 0 (black). Furthermore, a demonstration of the usage of the masks is shown in Figure 3.5, where they are used to compute the gradient of the central pixel in the patch.

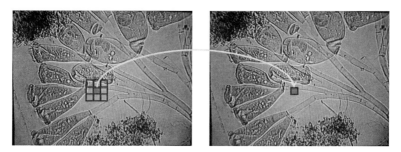

Figure 3.5: An example for the usage of mask matrix in Soble edge detection. The left image shows the working region of the masks in an image of *Vorticella*. The right image shows the objective pixel of the masks.

As the approach mentioned above, all edges in the image can be detected using an iterative strategy. An example using Sobel edge detection is shown in Figure 3.6.

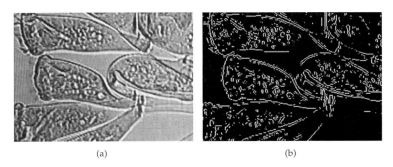

(a)   (b)

Figure 3.6: An example of Sobel edge detection. (a) An original image of *Vorticella*. (b) Edge detection result using the Sobel method.

Morphological Operations

Morphological operations are used as a tool for pre- and postprocessing of an image, including morphological filtering, opening, closing operations and so on [GW08]. Because morphological operations can enhance the properties of images, they are usually used together with edge detection techniques for the goal of image segmentation. Hence, several often used morphological operating techniques are introduced as follows:

- Median Filtering Operation:
  Before edge detection, median filter is usually used as a preprocessing for image denoising, leading to a higher performance of the image segmentation result [HYT79]. This filter uses the median value of an image patch to replace the original pixel value in the centre of this patch. By this filtering operation, the obvious different pixel values can be removed from the image. An example of median filter is shown in Figure 3.7.

- Opening Operation:
  Opening operation is used to break thin connections of regions in an image, which can smooth the edges of objects [GW08]. There are two main steps of this operation: First, erosion is used to break the thin connections of regions by decreasing their sizes. Then, dilation is applied to replenish the regions by increasing their sizes. A demonstration of the opening operation is shown in Figure 3.8.

- Closing Operation:
  Closing operation is used for linking broken connections of regions in an image,

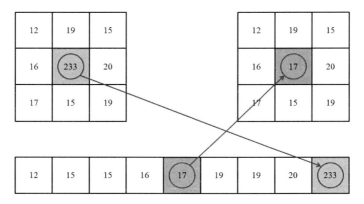

Figure 3.7: An example of a median filter for image denoising. The top left matrix shows the pixel values of an original image patch. The bottom array is the ranked pixel values. The top right matrix shows the pixel values of the filtered image patch. In the original patch, the central pixel (in the blue box) is a noisy pixel whose value is obviously higher than its surrounding pixels. By the filtering operation, the value of this pixel is replaced by the median value of all the pixels (in the red box.)

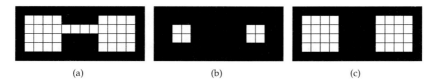

Figure 3.8: An example of an opening operation. (a) An original image. (b) The image after erosion. (c) The image after dilation.

which is useful for connecting independent objects [GW08]. Two phases are needed in this operation: The first is dilation, and the second is erosion. In Figure 3.9, an example of closing operation is shown.

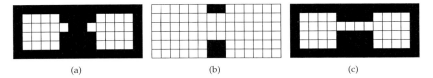

(a)   (b)   (c)

Figure 3.9: An example of a closing operation. (a) An original image. (b) The image after dilation. (c) The image after erosion.

- Filling Holes Operation:
  Filling holes operation is used to make up the blank holes in a closed edge [GW08]. After edge detection, this operation can obtain the whole inner region for image segmentation. An instance of this operation is shown in Figure 3.10.

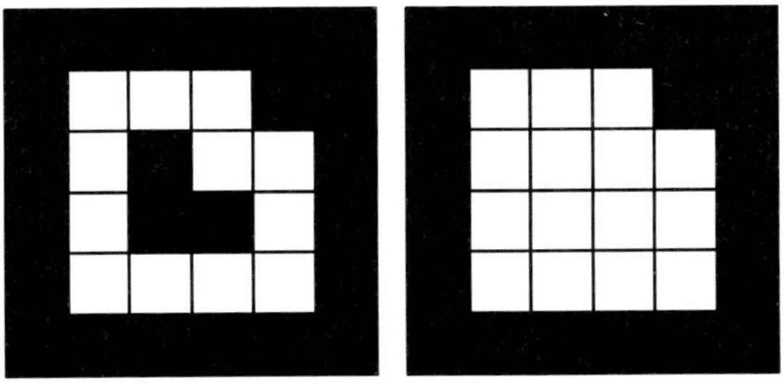

(a)   (b)

Figure 3.10: An example for a filling holes operation. (a) An original image with a blank hole in an object. (b) The image after filling holes operation.

3.2.2 Semi-automatic Image Segmentation

Motivation

Because microscopic images often contain a lot of uncertain and irregular styles of noise, they are hard to be segmented by full-automatic methods with a good performance. To this end, a Sobel edge detection based semi-automatic method is proposed to solve the segmentation problem in highly noisy images, e.g. the images of EMs in Figure 3.11(a).

Techniques

This proposed segmentation method has six main steps, including two manual operating steps. A demonstration of this method is shown in Figure 3.11. In Figure 3.11(a), an original image is put in. Then, an approximate region of the interesting object is chosen through manual cursor operation in Figure 3.11(b). Thirdly, Sobel edge detection is applied to search the edges of the object as shown in Figure 3.11(c). Fourthly, morphological operations are used to fill holes and smooth the edges in Figure 3.11(d). Fifthly, in Figure 3.11(e) the second manual operation is used to select the foreground depicting the object of interest. Finally, the segmentation result is obtained in Figure 3.11(f). The effectiveness of this method is shown in Section 7.1.2.

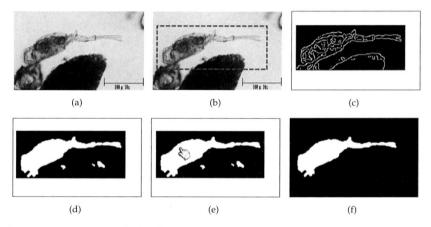

(a)	(b)	(c)
(d)	(e)	(f)

Figure 3.11: An example for the proposed semi-automatic segmentation approach. (a) An original microscopic image of *Rotifera*. (b) Manual region selection. (c) Sobel edge detection. (d) Morphological operations. (e) Manual object selection. (f) The final segmentation result.

3.3   Global Shape Features

In this section, the selected global shape feature extraction techniques are illuminated, including: isomerous edge histogram descriptor (IEHD), basic geometrical feature (BGF), Fourier descriptor (FD), and internal structure histogram (ISH). For all the mentioned features, the input data is a binary image (the segmentation result of Section 3.2), and the output data is a feature vector. Especially images of environmental microorganisms (EMs) are used as examples to explain the discriminative characteristics of these features, where the region of an EM (in white) is distinguished from the background (in black).

3.3.1   Isomerous Edge Histogram Descriptor

IEHD is a contour-based shape feature, which is developed to represent the continuous degree of edges in an image [LSG15a]. Firstly, as an example shown in Figure 3.12, the edges are used as the boundaries between the foreground and the background in the image .

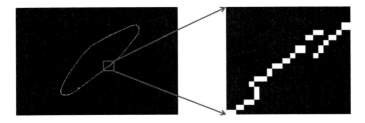

Figure 3.12: An example of the continuous degree of edges. The left image is the boundary of *Paramecium*. The right image shows an amplified part of the boundary.

Secondly, the number of pixels (edge length) in each edge (connected component) is counted. To count the number, 4-connected pixels are used as the touching neighbours of each pixel on the edges, where every pixel is connected horizontally and vertically [GW08]. Finally, these counted numbers are used to establish a histogram (the "isomerous edge histogram"). In this histogram, each bin contains the edges in a given interval of lengths, which expresses the distribution of edge lengths. In this work, a 13 bins histogram is defined with the interval of $[1,10]$, $[11,20]$, ..., $[111,120]$, $[120,+\infty)$ pixels, which leads to a 13-dimensional feature vector $c_{\text{IEHD}} = (\text{IEHD}_1, \text{IEHD}_2, \ldots, \text{IEHD}_{13})^{\text{T}}$, where $\text{IEHD}_{1,2,\ldots,13}$ represents the value of each bin, separately.

To show the discriminative property of IEHD, an instance is given in Figure 3.13, where the IEHDs of two EMs are compared. Because the *Actinophrys* in Figure 3.13(a) has many pseudopodium, it has a lot of dispersive edges and is assigned a large number in IEHD in Figure 3.13. In contrast, the *Arcella* in Figure 3.13(b) is very smooth, so the edges on its contour are very continuous, leading to a very small IEHD number as shown in Figure 3.13. Finally, these two EMs can be distinguished by their distinct IEHDs.

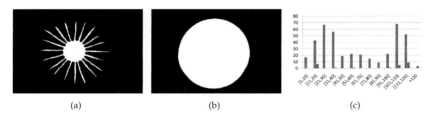

(a)   (b)   (c)

Figure 3.13: An example of IEHD. (a) An image of *Actinophrys*. (b) An image of *Arcella*. (c) IEHDs of (a) and (b). The blue bins represent the IEHD of (a), and the red bins show the IEHD of (b). The horizontal axis shows the 13 intervals of edges. The vertical axis shows the amount of edges.

3.3.2 Basic Geometrical Features

BGF contains the fundamental geometrical measurements of objects, which is a region-based shape feature. In the following part, 23 elements of BGF are explained. For an image with the size of $M \times N$ pixels in particular, the pixel value at the position (x, y) is defined as $f(x, y)$.

Perimeter

The perimeter of an object is the total length of its contour, which is the sum of all pixel values of all edges. The definition of perimeter BGF_1 is as follows

$$BGF_1 = \sum_{x=1}^{M} \sum_{y=1}^{N} f(x, y) \quad , \quad \text{if} \quad (x, y) \quad \text{is on the edges} \quad . \tag{3.6}$$

Area

The area of an object is the size of its body, which the sum of all the pixel values in the object. The equation below defines area BGF_2 as

$$BGF_2 = \sum_{x=1}^{M} \sum_{y=1}^{N} f(x,y) \quad , \quad \text{if} \quad (x,y) \quad \text{is in the object} \quad . \tag{3.7}$$

Complexity

Complexity describes the complex level (or complex rate) of an object, which is defined as

$$BGF_3 = \frac{BGF_1^2}{4\pi BGF_2} \quad . \tag{3.8}$$

From Eq. 3.8, people can find that when an object has a circular shape, its complexity has the lowest value of 1. When the shape changes to a more complex format, its complexity increases. An example of complexity is shown in Figure 3.14, where the complexities of four EMs are compared. The first *Actinophrys* has a very complex radial shape, so it has the highest value of complexity. The second *Aspidisca*'s main body is a simple circular shape, but since it has a structure of cilia, its complexity is increased. Although the third *Paramecium* is big, its shape is a simple ellipse, so its complexity is low. The last *Arcella* has a perfectly circular shape, so its complexity is very near to 1.

| 82.50 | 3.51 | 2.11 | 1.02 |

Figure 3.14: An example of the complexity. From left to right, four EM images show *Actinophrys*, *Aspidisca*, *Paramecium* and *Arcella*, respectively. The value under each image is the complexity of the above mentioned EM.

Long Side of Bounding Box

The long side of bounding box of an object is an approximate length for describing the major axial length of the object. Because the bounding box is along the horizontal and vertical directions, an object rotation preprocessing is needed before the feature extraction. To this end, a Hotelling transformation (also known as Karhunen-Loeve transform, eigenvector transform and principal component analysis), is used to unify the directions of shapes [Jol02] which can rotate an object into the direction of its major axis. This transformation is defined in the following

$$
\begin{aligned}
\mathbf{m}_{(x,y)_i} &= \frac{1}{\text{BGF}_1} \sum_{i=1}^{\text{BGF}_1} (x,y)_i \quad , \\
\mathbf{c}_{(x,y)_i} &= \frac{1}{\text{BGF}_1} \sum_{i=1}^{\text{BGF}_1} [(x,y)_i - m_{(x,y)_i}][(x,y)_i - m_{(x,y)_i}]^{\text{T}} \quad ,
\end{aligned}
\tag{3.9}
$$

where $(x,y)_i$ is the coordinate of the ith pixel on the contour of the shape, $\mathbf{m}_{(x,y)_i}$ is the mean vector of $(x,y)_i$, and $\mathbf{c}_{(x,y)_i}$ is the covariance matrix of $(x,y)_i$. Because $\mathbf{c}_{(x,y)_i}$ is a 2-by-2 symmetric matrix, it only has two eigenvectors. Using these two eigenvectors, a new matrix can be constructed as $\mathbf{k} = (\mathbf{k}_1, \mathbf{k}_2)^{\text{T}}$, where $\mathbf{k}_1$ is the eigenvector corresponding to the larger eigenvalue, and $\mathbf{k}_2$ is the eigenvector corresponding to the smaller eigenvalue. Based on this matrix $\mathbf{k}$, a set of new vectors can be calculated as follows

$$
(x,y)_{i,\text{new}} = \mathbf{k} \times [(x,y)_i - \mathbf{m}_{(x,y)_i}] \quad , \quad (i = 1,2,3,\ldots,\text{BGF}_1) \quad .
\tag{3.10}
$$

Using the obtained vectors from Eq. 3.10, a new coordinate system can be first built up which uses the position of $\mathbf{m}_{(x,y)_i}$ as the origin, the direction of $\mathbf{k}_1$ as the orientation of the new horizontal axis, and the direction of $\mathbf{k}_2$ as the orientation of the new vertical axis. Then, the shape is transformed from the original coordinate system to the new one. Finally, all the shapes in different directions can be unified into the same main direction by this Hotelling transformation. An example of Hotelling transformation on image rotation is shown in Figure 3.15.

Based on the rotation technique mentioned above, the long side of bounding box of an object can be defined as follows

$$
\text{BGF}_4 = x_{\text{rightest,new}} - x_{\text{leftest,new}} \quad ,
\tag{3.11}
$$

where $x_{\text{rightest,new}}$ is the rightest position of the contour of the object in the new coordinate system, and $x_{\text{leftest,new}}$ is the leftest position of the contour of the object in the new coordinate system.

Short Side of Bounding Box

Similar to the definition of the long side of bounding box BGF_4, the short side of bounding box is defined as follows

$$
\text{BGF}_5 = y_{\text{highest,new}} - y_{\text{lowest,new}} \quad ,
\tag{3.12}
$$

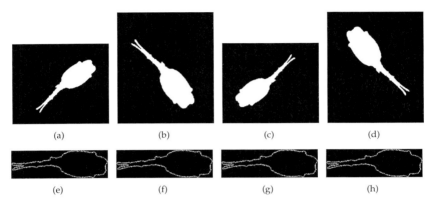

(a)   (b)   (c)   (d)

(e)   (f)   (g)   (h)

Figure 3.15: An example of Hotelling transformation for image rotation. (a) is the original image of *Rotifera*. (b), (c), and (d) are the rotated results of (a) with 90°, 180°, and 270°, separately. (e), (f), (g), and (h) are the rotation results of the Hotelling transformation of (a), (b), (c), and (d), respectively.

where $y_{highest,new}$ is the highest position of the contour of the object in the new coordinate system, and $y_{lowest,new}$ is the lowest position of the contour of the object in the new coordinate system.

Elongation of Bounding Box

Elongation of bounding box describes the length-width ratio of an object and it is defined as follows

$$BGF_6 = \frac{BGF_4}{BGF_5} \quad , \tag{3.13}$$

where, when the object is "long" and "thin", BGF_6 is larger, and if the object is "short" and "fat", BGF_6 is smaller.

Major Axis Length

The length of the major axis of an object is a measure of the length of the object, which uses an ellipse to approximately represent the object, and uses the length of the long axis of this ellipse to describe the length of the object. To obtain the approximate ellipse, there are many approaches, and the ellipse that has the same normalized second central moments as the object is used in this work. The major axis length is defined in the following

$$BGF_7 = \sum_{x=1}^{M} \sum_{y=1}^{N} f(x,y) \quad , \quad \text{if} \quad (x,y) \quad \text{is on the long axis of the approximate ellipse} \quad .$$

$$(3.14)$$

Minor Axis Length

Similar to the definition of BGF_7, the length of the minor axis of an object describes the width of the object, using the length of the short axis of the approximate ellipse to represent its width. The minor axis length is defined as follows

$$BGF_8 = \sum_{x=1}^{M} \sum_{y=1}^{N} f(x,y) \quad , \quad \text{if} \quad (x,y) \quad \text{is on the short axis of the approximate ellipse} \quad .$$

$$(3.15)$$

Elongation of Approximate Ellipse

The elongation of the approximate ellipse represents the length-width ratio of an object, and it is defined as

$$BGF_9 = \frac{BGF_7}{BGF_8} \quad , \tag{3.16}$$

where, if the object is "short" and "fat", BGF_9 is smaller, but if the object is "long" and "thin", BGF_9 is larger.

Eccentricity

Based on the approximate ellipse of an object, the eccentricity is defined as

$$BGF_{10} = \left(1 - \frac{BGF_8^2}{BGF_7^2}\right)^{1/2} \quad , \tag{3.17}$$

where the eccentricity BGF_{10} is the ratio of the distance between the focus of the ellipse and its long axis length. The value of BGF_{10} is in $[0,1]$. When the shape of the object is close to a circle, its BGF_{10} approaches 0. In contrast, when the shape is "long" and "narrow", the BGF_{10} is close to 1.

Equiv Diameter

The equiv diameter of an object is defined as

$$\text{BGF}_{11} = \left(4\frac{\text{BGF}_2}{\pi}\right)^{1/2} = 2\left(\frac{\text{BGF}_2}{\pi}\right)^{1/2} \quad , \qquad (3.18)$$

where BGF_{11} represents the length of the diameter of a circle which has the same area BGF_2 as the object.

Surface Roughness

Surface roughness uses the variance of centre distances of an object's edges to measure the smooth degree of the contour of the object. It is defined as follows

$$d_{\text{mean}} = \frac{1}{\text{BGF}_1} \sum_{i=1}^{\text{BGF}_1} [|(x,y)_i - \mathbf{m}_{(x,y)_i}|] \quad ,$$

$$\text{BGF}_{12} = \frac{1}{\text{BGF}_1} \sum_{i=1}^{\text{BGF}_1} [|(x,y)_i - \mathbf{m}_{(x,y)_i}| - d_{\text{mean}}]^2 \quad , \qquad (3.19)$$

where $(x,y)_i$ is the coordinate of the ith pixel on the contour of the shape, $\mathbf{m}_{(x,y)_i}$ is the mean vector of $(x,y)_i$ (defined in Eq. 3.10), d_{mean} is the mean of centre distances of the contour of the object, and BGF_{12} is the finally obtained surface roughness. When the object's surface is smooth, BGF_{12} approaches 0. When the the surface is rougher, the value of BGF_{12} becomes higher.

Mean Value of Boundary Curvature

The mean value of boundary curvature of an object is defined as follows

$$\text{BGF}_{13} = \frac{1}{\text{BGF}_1} \sum_{i=1}^{\text{BGF}_1} \frac{|x_i' y_i'' - x_i'' y_i'|}{[(x_i')^2 + (y_i')^2]^{3/2}} \quad , \qquad (3.20)$$

where x_i is the x element of $(x,y)_i$, y_i is the y element of $(x,y)_i$. BGF_{13} describes the degree of the bend of the integral object. If the object has more curves on its contour, the value of BGF_{13} is higher. If the object has more straight edges, the BGF_{13} is lower.

Variance Value of Boundary Curvature

Based on BGF_{13} in Eq. 3.20, the variance value of boundary curvature is defined as

$$\text{BGF}_{14} = \frac{1}{\text{BGF}_1} \sum_{i=1}^{\text{BGF}_1} \left\{ \frac{|x_i' y_i'' - x_i'' y_i'|}{[(x_i')^2 + (y_i')^2]^{3/2}} - \text{BGF}_{13} \right\}^2 \quad . \qquad (3.21)$$

BGF_{14} is a measurement of the stability of the object's contour, where when it is larger, its contour is combined by more categories of curves.

First Order Moment

The first order moment of an object is the mean of distances between the pixels in the object and its geometrical centre. This moment measures the intensity of the pixels in the object, where when two objects have the same amounts of pixels, if it is larger, the distribution of pixels is more sparse, otherwise the distribution of pixels is more concentrated. This moment is defined in the following

$$\mathbf{m}_{(x,y)_j} = \frac{1}{BGF_2} \sum_{j=1}^{BGF_2} (x,y)_j \quad ,$$

$$BGF_{15} = \frac{1}{BGF_2} \sum_{j=1}^{BGF_2} [|(x,y)_j - \mathbf{m}_{(x,y)_j}|] \quad ,$$

(3.22)

where $(x,y)_j$ is the coordinate of the jth pixel in the object, $\mathbf{m}_{(x,y)_j}$ is the mean vector (the position of the geometrical centre of the object) of $(x,y)_j$, and BGF_{15} is the first order moment of this object.

Second Order Moment

The second order moment of an object is the variance of distances between the pixels in the object and its geometrical centre, which measures the stability of these distances. When this moment is close to 0, it means that these distances have a stable distribution. The definition of this moment is given as follows

$$BGF_{16} = \frac{1}{BGF_2} \sum_{j=1}^{BGF_2} [|(x,y)_j - \mathbf{m}_{(x,y)_j}| - BGF_{15}]^2 \quad ,$$

(3.23)

where $(x,y)_j$ is the coordinate of the jth pixel in the object, $\mathbf{m}_{(x,y)_j}$ is the mean vector of $(x,y)_j$ (defined in Eq. 3.22), and BGF_{16} is the second order moment of this object.

Third Order Moment

The third order moment of an object is the skewness of distances between the pixels in the object and its geometrical centre, which is a measurement of the bias of the distribution of pixels. When this moment is negative, the distribution biases to the left of the centre. Otherwise, when this moment is positive, the distribution biases to the right of the centre. The third order moment is defined as follows

$$BGF_{17} = \frac{\sum\limits_{j=1}^{BGF_2} [|(x,y)_j - \mathbf{m}_{(x,y)_j}| - BGF_{15}]^3}{BGF_2 BGF_{16}^{3/2}} \quad . \quad (3.24)$$

Fourth Order Moment

The fourth order moment of an object is the kurtosis of distances between the pixels in the object and its geometrical centre, which measures the stability of the variance BGF_{16}. When this moment is larger, it indicates that BGF_{16} is caused by distributions with a low frequency. Furthermore, these distributions can be considered as noisy parts of the object. This moment is defined as

$$BGF_{18} = \frac{\sum\limits_{j=1}^{BGF_2} [|(x,y)_j - \mathbf{m}_{(x,y)_j}| - BGF_{15}]^4}{BGF_2 BGF_{16}^2} - 3 \quad . \quad (3.25)$$

Median of Centre Distances

The median of centre distances BGF_{19} is a normal statistical measure in geometry, which separates the higher half from the lower half of all the distances.

Maximal Value of Centre Distances

The maximal value of centre distances BGF_{20} is used to evaluate the case of extreme (maximum value), where the biggest centre distance is defined as this value.

Minimal Value of Centre Distances

The maximal value of centre distances BGF_{21} is a measure of the case of extreme (minimal value), which is defined as the smallest centre distance of an object.

Range Value of Centre Distances

The range value of centre distances of an object describes the changing scale of these distances, which is defined in the following

$$BGF_{22} = BGF_{20} - BGF_{21} \quad . \quad (3.26)$$

Mode Value of Centre Distances

The mode value of centre distances of an object BGF_{23} is a distance which occurs the most often among all these distances.

Finally, all of the 23 components mentioned of BGF constitute a 23-dimensional feature vector $c_{BGF} = (BGF_1, BGF_2, \dots, BGF_{23})^T$. To use c_{BGF}, it is not necessary to apply all 23 dimensions at the same time, but better to select the suitable ones for different tasks.

3.3.3   Fourier Descriptor

Shape Signature

SS represents the shape of an object by a one-dimensional function that is derived from shape boundary points [ZL02; GS07]. There are many types of SS, and a centroid distance based SS is used in this work because of its generality in different tasks. This SS computes Euclidean distances $d_{t=1,\dots,K}$ from a shape centre to K points equidistantly distributed along its contour. To demonstrate the discriminative function of the SS, an example is given in Figure 3.16, where different objects have obvious differences on their SSs.

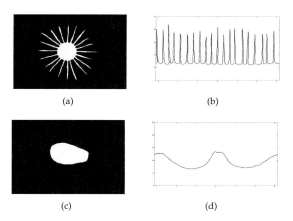

(a)   (b)

(c)   (d)

Figure 3.16: An example of the discriminative power of SS. (a) and (c) are the images of *Actinophrys* and *Euglypha*, respectively. (b) and (d) are the SSs of (a) and (c), separately. The horizontal axis in (b) and (d) shows the points on the contour of each object, and the vertical axis shows the centriod distances of these two objects, respectively.

Although SS can accurately describe the shape characteristic of the object, it is not robust to noise, rotation or scaling. For example, Figure 3.17 shows an example of the object rotation problem of SS, where the same object has two different SSs when it locates in different directions.

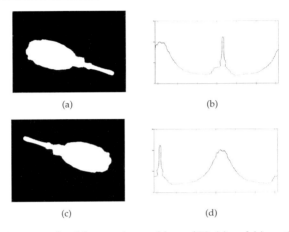

(a)   (b)

(c)   (d)

Figure 3.17: An example of the rotation problem of SS. (a) and (c) are the original and rotated (180°) images of *Rotifera*, respectively. (b) and (d) are the SSs of (a) and (c), separately. The setting of horizontal and vertical axis in (b) and (d) is the same as that in Figure 3.16.

To solve the noise problem, the median filtering operation described in Section 3.2.1 is used to remove the noise. In order to overcome the problem of object rotation, the Hotelling transformation introduced in Section 3.3.2 is applied to unify the directions of objects. To resolve the scaling problem, a normalisation is used to the extracted SS, where all the distances are mapped into an interval of $[0,1]$. Finally, in this work $K = 100$ sample points on the contour of each object are used based on pretests, leading to a 100-dimensional feature vector $c_{SS} = (d_1, d_2, \ldots, d_K)^T$.

Fourier Descriptor

To solve the noise, rotation and scaling problems of SS, FD is an effective solution [ZL03]. FD uses Fourier transform to represent the overall shape of an object by the first few low-frequency terms and shape details by higher frequency-terms.

Specifically, the discrete Fourier transform on the centroid distance d_t is defined as

$$a_k = \frac{1}{K} \sum_{t=0}^{K-1} d_t \exp\left(\frac{-j2\pi kt}{K}\right) \quad . \tag{3.27}$$

The resulting coefficients a_k $(k = 0, \ldots, K-1)$ are still sensitive to noise, rotation and scaling. To overcome this, they are further normalised as

$$b_k = \frac{a_k}{a_0} \quad . \tag{3.28}$$

Due to the symmetry $b_k = b_{K-1-k}$, the dimensionality of the resulting feature vector is $K/2$. As a result, FD is represented by a $K/2$-dimensional vector $\mathbf{c}_{FD} = (b_0, \ldots, b_{K/2-1})^{T}$ ($K = 50$ is used in the experiments of this work).

3.3.4 Internal Structure Histogram

To capture the structure of an object, ISH first computes the internal structure angles (ISAs) $\angle\beta_{u,v,w}$ based on its contour as shown in Figure 3.18(a). For this, K sample points ($K = 100$ in the experiments of this work) are equidistantly distributed along the contour of the object and inscribed angles are created for all possible combinations of point triples among them. Then, a histogram is built by classifying angle values into 10 bins defined as $[0°, 36°), [36°, 72°), \ldots, [324°, 360°)$. Thus, the resulting ISH is represented as a 10-dimensional feature vector $\mathbf{c}_{ISH} = (ISH_1, ISH_2, \ldots, ISH_{10})^{T}$, where $ISH_{1,2,\ldots,10}$ is the value of each bin, respectively.

By referring to Figure 3.18, discriminative properties of ISH are explained. Figure 3.18(b) and Figure 3.18(c) demonstrate the robustness of ISH to scaling and rotation. Specifically, even though an object is scaled or rotated, sample points are located at the same positions on its contour. Thus, ISAs obtained from the rotated or scaled object are the same as those obtained from the original object. Hence, ISHs consisting of these ISAs are the same among the original, rotated and scaled cases. However, ISH is sensitive to shape deformation. Since the deformed object contains ISAs that are dissimilar to the ones obtained form the original object, ISHs in these cases are different. For instance, in Figure 3.18(d), the tail of the EM bends to the left. This deformation leads to a histogram which is different from the one in Figure 3.18(a). However, this disadvantage can be alleviated by fusing ISH with features that are robust to shape deformation, such as BGF considering the perimeter (BGF_1) and area (BGF_2) of an object.

It should be noted that the ISH deals with the frequencies (statistics) of angles defined on the object boundary. With respect to this, one ISA $\angle\beta_{u,v,w}$ is related to the other two ISAs $\angle\beta_{u,w,v}$ and $\angle\beta_{w,u,v}$ (i.e., $\angle\beta_{u,v,w} = 180° - (\angle\beta_{u,w,v} + \angle\beta_{w,u,v})$). However,

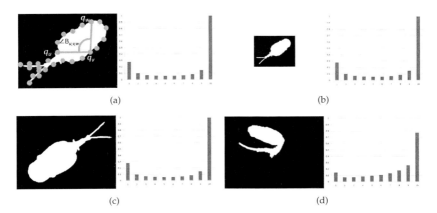

Figure 3.18: An illustration of discriminative properties of ISH. The original image of *Rotifera* in (a) is scaled, rotated and deformed in (b), (c) and (d), respectively.

an ISH created by two of such three ISAs loses the shape information of the object. This is because the relation of each sample point to all the others is needed to fully represent the shape of the object. Hence, an ISH that preserves frequencies of all the ISAs is created in this work.

3.4   Local Shape Features

Different from the aforementioned features (in Section 3.3) that describe the global shape characteristics of an object, local shape features focus on the regional shape properties of the object. Above all, the SIFT feature is selected in this work, because of its effectiveness in image description.

The SIFT feature is extracted from original images that have no image segmentation. This process consists of three steps [Jia+10]: Given an image, the first step, region detection, detects regions that are useful for characterising objects. The second step is region description where each of the detected regions is quantified as a vector, that is the SIFT feature. In the last aggregation step, the SIFT feature extracted from the image is aggregated into a histogram representation.

For the region detection step, a preliminary experiment shows that region detectors (e.g. Harris-Affine and Hessian-Affine detectors [Mik+05]) which select characteristic regions in an image, do not work well on microscopic images because of their insufficient contrasts. Hence, the dense sampling technique is chosen, for example regions with the radius of four pixels are exhaustively detected with an interval of every six pixels. In the region description step, each region is represented as a 128-dimensional SIFT feature vector which describes the distribution of edge orientations in that region. Finally, the aggregation step is based on the BoVW approach which represents an image as a collection of characteristic SIFT features, namely *Visual Words* [Jia+10]. Roughly speaking, one million of SIFT features are randomly selected from the image datasets in this work, and k-means clustering is used to group them into 1000 clusters, where each cluster centre is a visual word. Then, by assigning the SIFT feature extracted from an image to the most similar visual words, this image is represented as a 1000-dimensional feature vector $\mathbf{c}_{\text{SIFT}} = (\text{SIFT}_1, \text{SIFT}_2, \ldots, \text{SIFT}_{1000})^{\text{T}}$, where each dimension represents the frequency of a visual word.

3.5   Strongly Supervised Learning

Strongly supervised learning (SSL) means that the training process of a classifier does not only need labels (or categorical information) of the input data, but also needs additional annotations of the data. For example, if an image classification process uses segmented images as training data, it is a strongly supervised learning approach. The RBFSVM classifier in particular is selected and applied in this work, because of its generality and high efficiency.

3.5.1   Fundamentals of SVM

For microscopic image classification, an SVM which extracts a decision boundary between images of different EM classes based on the margin maximisation principle is selected. Due to this principle, the generalisation error of the SVM is theoretically independent of the number of feature dimensions [Vap98]. Furthermore, a complex decision boundary can be extracted using a non-linear SVM.

Suppose a labelled training set of H images represented by feature vectors $\mathbf{c}_{i=1,2,\ldots,H}$. A two-class classification problem is considered, where positive images showing the interesting objects of a certain category are associated with the label $l_i = 1$, and negative ones showing objects of the other categories with $l_i = -1$. In the training process of an SVM, a hyperplane is placed in the middle between positive and negative images, so that the distance (margin) of the hyperplane to the nearest positive (or negative) image is maximised. Mathematically, the hyperplane is determined by maximising the following objective function:

$$\hat{\mathbf{w}} = \underset{\mathbf{w}}{\operatorname{argmax}} \left[\sum_{i=1}^{H} w_i - \frac{1}{2} \sum_{i=1}^{H} \sum_{j=1}^{H} w_i w_j l_i l_j K(\mathbf{c}_i, \mathbf{c}_j) \right] \quad , \quad \text{where}$$

$$\mathbf{w} = (w_1, w_2, \ldots, w_i, \ldots, w_H)^{\mathrm{T}} \quad ; \quad 0 \le w_i \le R \quad ; \quad \sum_{i=1}^{H} w_i l_i = 0 \quad .$$

(3.29)

w_i is the weight associated with $\mathbf{c}_i$, $R > 0$ is the penalty parameter of the classification error. The kernel value $K(\mathbf{c}_i, \mathbf{c}_j)$ represents the product between $\mathbf{c}_i$ and $\mathbf{c}_j$, which are mapped into a high-dimensional space. The SVM finds a linear hyperplane with the maximum margin in the higher dimensional feature space. For example, when the SVM has a kernel function as follows

$$K(\mathbf{c}_i, \mathbf{c}_j) = \mathbf{c}_i^{\mathrm{T}} \mathbf{c}_j \quad , \quad (3.30)$$

it is a standard linear SVM, which is good at classifying high dimensional and large scale data with a low computational complexity. A visible demonstration of the linear SVM is shown in Figure 3.19(a), where a linear SVM classifier is trained by the "green circles" and "red triangles".

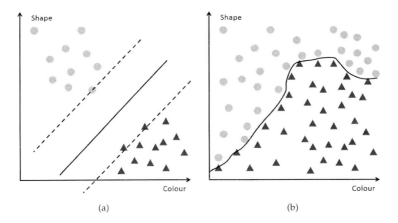

(a)   (b)

Figure 3.19: An example of SVMs. (a) is a linear SVM. (b) shows an RBFSVM.

After constructing the SVM, a test image represented by $\mathbf{c}$ is classified using the following function

$$\xi(\mathbf{c}) = \sum_{i=1}^{H} \hat{w}_i l_i K(\mathbf{c}_i, \mathbf{c}) - b \quad , \quad (3.31)$$

where b is the bias of the hyperplane computed by taking the derivative of the maximisation term in Eq. (3.29) with any $w_i > 0$. $\xi(\mathbf{c})$ represents the distance between $\mathbf{c}$ and the separating hyperplane of the SVM. The larger $\xi(\mathbf{c})$ is, the more likely $\mathbf{c}$ represents the relevant category of the image.

3.5.2 Radial Basis Function Kernel SVM

Among many existing kernels of the SVM, RBF has an outstanding generality on different classification tasks [CL11]. It is defined as follows

$$K(\mathbf{c}_i, \mathbf{c}_j) = \exp^{-\gamma \|\mathbf{c}_i - \mathbf{c}_j\|^2} \quad \text{with} \quad \gamma > 0 \quad , \tag{3.32}$$

where γ is a kernel parameter and refers to the curvature of the SVM margin. By adjusting the original straight hyperplane to be curving, the classification accuracy is improved. In order to use the above mentioned RBF kernel for SVM, the penalty value R and the kernel parameter γ need to be adjusted. To this end, this work heuristically sets $R = 2$ and γ to the average value of squared Euclidean distances of all pairs of feature vectors of microscopic images. A visible instance of RBFSVM is shown in Figure 3.19(b), where an RBFSVM classifier is trained by the "green circles" and "red triangles" as a complex curve.

3.5.3 Late Fusion of SVM

In many image classification tasks, using only single features is not enough to describe the characteristics of the images completely. Hence, a late fusion approach is selected in this work, due to its robust performance mentioned in Section 2.1.4. Especially because SVMs are used as classifiers in this dissertation, the late fusion that works on SVMs is discussed in the following.

In order to simplify the fusion of SVM outcomes obtained for different features, the values of $\xi(\mathbf{c})$ are transformed into the range $(0,1)$ by applying a sigmoid function [LLW07]

$$\Xi(\mathbf{c}) = \sigma(\xi(\mathbf{c})) \in (0,1) \quad \text{with} \quad \sigma : \mathbb{R} \to (0,1) \quad . \tag{3.33}$$

This $\Xi(\mathbf{c})$ is called 'probabilistic outputs for SVMs'. Since this sigmoid function is monotonically increasing, a value calculated with Eq. (3.33) corresponds to the likelihood of $\mathbf{c}$ to represent an image of the relevant (correct) category.

In order to classify an image using late fusion, first, its feature vectors are extracted with the methods introduced in Section 3.3. In this way, each image is represented by a set C of different features as below

$$C = \{\mathbf{c}^1, \quad \mathbf{c}^2, \quad \mathbf{c}^3, \quad \ldots, \quad \mathbf{c}^{|C|}\} \quad . \tag{3.34}$$

where $|C|$ is the number of the applied features.

Second, probabilistic outputs are computed (according to Eq. (3.33)) separately for all the feature representations

$$
\begin{array}{ccccc}
C = & \{\mathbf{c}^1, & \mathbf{c}^2, & \mathbf{c}^3, & \ldots, & \mathbf{c}^{|C|}\} \\
& \downarrow & \downarrow & \downarrow & \downarrow & \downarrow \\
& \xi(\mathbf{c}^1) & \xi(\mathbf{c}^2) & \xi(\mathbf{c}^3) & \xi(\mathbf{c}^{\cdots}) & \xi(\mathbf{c}^{|C|}) \\
& \downarrow & \downarrow & \downarrow & \downarrow & \downarrow \\
& \Xi(\mathbf{c}^1) & \Xi(\mathbf{c}^2) & \Xi(\mathbf{c}^3) & \Xi(\mathbf{c}^{\cdots}) & \Xi(\mathbf{c}^{|C|})
\end{array}
\tag{3.35}
$$

Further, these probabilistic outputs are fused by a linear combination

$$
\theta(C) = \sum_{i=1}^{|C|} \lambda^i \Xi(\mathbf{c}^i) \quad \text{with} \quad \lambda^i \in [0,1] \quad \text{and} \quad \sum_{i=1}^{|C|} \lambda^i = 1 \quad, \tag{3.36}
$$

which provides an overall assessment of the image described by C. The weight λ^i indicates the usefulness of the corresponding feature vector $\mathbf{c}^i$.

To obtain such weights λ^i, there are many feasible methods, and a grid search strategy is applied in this work, because it is simple and effective. First, the set of training images is divided into two subsets with the same cardinality. Then, one of these two subsets is used for training (building SVMs) and the other one is used for testing (validation). Subsequently, all possible combinations of weights are examined using the grid search strategy. Each weight is quantised into 21 values with the step length 0.05, that is $\{0, 0.05, 0.1, 0.15, \ldots, 1\}$. The weigh combination which leads to the highest average precision (AP) is selected as the final useful λ^i in Eq. 3.36. AP has been developed in the field of information retrieval, and is used as an indicator to evaluate a ranked list of retrieved samples [Kis05]. Assume that given a weight combination, T images are ranked based on their likelihoods (probabilistic outputs) $\theta(C)$. Under this setting, AP is defined as follows:

$$
\text{AP} = \frac{\sum_{h=1}^{T} [P(h) \times rel(h)]}{\text{number of relevant images}} \quad, \tag{3.37}
$$

where $P(h)$ is the precision by regarding a cut-off position as the h-th position in the list, and $rel(h)$ is an indicator function which takes 1 if the image ranked at the h-th position is relevant, otherwise 0. Thus, AP represents the average of precisions each of which is computed at the position where a relevant image is ranked. The value of AP increases if images relevant to a target image category are ranked at higher positions in the image list sorted based on evaluation values. Therefore, AP can be used to measure the classification performance for each of the weight combinations. Finally, the weight combination with the highest AP is used to fuse SVMs which are built using all training images.

3.6   Summary

In this chapter, the proposed semi-automatic microscopic image classification system using SSL framework is introduced. First, a semi-automatic image segmentation method is stated in Section 3.2. Then, selected global shape features are introduced in Section 3.3. Thirdly, a local shape feature is proposed in Section 3.4 Finally, SSL classifiers and late fusion methods are proposed in Section 3.5.

Chapter 4

Full-automatic Microscopic Image Classification Using Weakly Supervised Learning

In this chapter, a full-automatic microscopic image classification system is built using a weakly supervised learning (WSL) framework. An overview of this system is introduced in Section 4.1 first. Then, sparse coding (SC) features are proposed in Section 4.2. Thirdly, the WSL classifier is presented in Section 4.3. Finally, this chapter is briefly concluded by Section 4.4. Especially environmental microorganism (EM) images are used as examples in this chapter to explain the technical approaches.

4.1 Overview of the System

In this chapter, a WSL system (see Section 4.3) which conducts EM classification by directly analysing microscopic images is developed. This EM classification task is considered as a binary classification problem using two types of training images: positive images where a certain EM is present, and negative images where it is absent. By comparing these training images, a classifier is built to discriminate between test images where the EM is present and the other test images. This EM classification system is developed by addressing the following two problems:

- Small Training Dataset:
  Environmental investigations are always operated in outdoor environments, where conditions like temperature and salinity are changing continuously. Because EMs are very sensitive to these conditions, their quantity is easily influenced. Thus, it is difficult to collect a large number of EM images, leading to

a small training dataset problem. To this end, SC features are applied to represent EM images because they can extract more efficient features to resolve the shortage of training images (see Section 4.2).

- Noisy Images:
  The majority of EM samples are obtained from complex environments, where a large amount of impurities like rubbish is present (see Figure 1.2(a)). The noise degrades the performance of EM classification, leading to a noisy image problem. To overcome this, WSL is applied, which is useful for excluding noise in surrounding regions of an EM and identifying its class (see Section 4.3).

To jointly solve the problems above, an EM classification system which incorporates SC features into a WSL framework is proposed. In this framework, SC features are extracted from EM images and used to train an improved RBSVM classifier (see Section 4.3.2). Figure 4.1 illustrates an overview of this system. In (a), two classes of weakly labeled training EM images are given. In (b), the training images are represented by SC features. In (c), an improved RBSVM classifier is built by iteratively finding the region of the most interesting EM in each training image, and updating the classifier using newly found regions. In (d), the finally obtained classifier is used to localise and classify the interesting EM in test images. There are two main contributions of this work: First, a full-automatic EM classification system using a WSL framework is developed. Second, the basic RBSVM algorithm is significantly improved using SC features.

4.2   Sparse Coding Features

Sparse coding (SC) feature can represent a vector by a sparse linear combination of an over-complete set of basis functions, which is effective to convert raw pixel values into higher-level semantic features for describing abstract characteristics of an image. In contrast to the hand-crafted (or predefined) features mentioned in Section 3.3 and 3.4, SC feature only uses the original images as input data and extracts feature vectors directly. In this section, the fundamental of SC is first introduced in Section 4.2.1. Then, a non-negative sparse coding (NNSC) technique is discussed in Section 4.2.2.

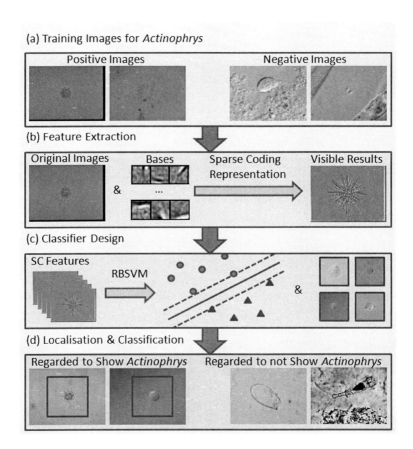

Figure 4.1: An overview of the EM classification system using WSL framework.

4.2.1   Sparse Coding

Basis and Linear Combination

Assume that $\mathbf{s}$ is a row vector in an S-dimensional linear space. It can be represented as a linear combination of S linearly independent elements $\{\mathbf{s}_1, \mathbf{s}_2, \ldots, \mathbf{s}_S\}$ as follows

$$\mathbf{s} = g_1\mathbf{s}_1 + g_2\mathbf{s}_2 + \ldots + g_S\mathbf{s}_S \quad , \tag{4.1}$$

where $g_{1,2,\ldots,S}$ is the weight related to $\mathbf{s}_{1,2,\ldots,S}$, separately. Furthermore the set of $\mathbf{s}_{1,2,\ldots,S}$ is called a basis set of $\mathbf{s}$. Based on this theory of linear algebra, an image could be represented by a linear combination of a basis set, too. First, the matrix of the image is converted into the format of the vector $\mathbf{s}$. Then this vector is represented by the linear combination of its bases $\mathbf{s}_{1,2,\ldots,S}$. An example of this image reconstructing process is shown in Figure 4.2.

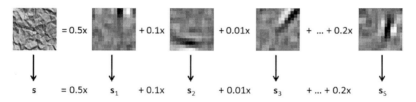

Figure 4.2: An example of image reconstruction by the linear combination of bases.

In Eq. 4.1, if the set of basis is very large, which covers all cases of basis, it is an over-complete set of basis elements, an example of which is shown in Figure 4.3. When the image is reconstructed by the over-complete basis set, many bases are not used and only have weights of 0, leading to a sparse weight vector like $(0.1, 0, 0, 0, 0, 0, 0.02, 0, 0, 0, 0, \ldots, 0, 0, 0.15, 0, 0, 0.01)$. This sparse vector of the weights is effective to extract semantic information from the image, which is called sparse coding (or sparse representation).

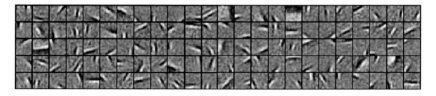

Figure 4.3: An example of the over-complete set of basis elements.

Sparse Coding for Image Representation

Based on the fundamental mentioned in Section 4.2.1, an image can be represented by SC approaches. Assume that an image patch $\mathbf{p}_i$ is converted into a Q-dimensional row vector which represents the brightness of each pixel, and $\mathbf{P}$ is a $P \times Q$ matrix consisting of P patches, where the ith row shows the ith path $\mathbf{p}_i$. $\mathbf{b}_j$ is a Q-dimensional row vector representing a basis, and $\mathbf{B}$ is a $G \times Q$ matrix representing a set of G bases, where the jth row shows the jth basis $\mathbf{b}_j$. SC aims to reconstruct $\mathbf{p}_i$ by a sparse linear combination of G bases in $\mathbf{B}$. To combine these G bases, a weight is assigned to each $\mathbf{b}_j$, and $\mathbf{a}$ is a G-dimensional row vector representing the set of weights. This reconstruction is solved by the following optimisation:

$$\underset{\mathbf{a},\mathbf{B}}{\text{minimise}} \quad \sum_{i=1}^{P} \| \mathbf{p}_i - \mathbf{a}_i \mathbf{B} \| + \mu \sum_{i}^{P} \mathbf{a}_i^j \quad , \tag{4.2}$$

where $\mathbf{a}_i$ is the weight vector corresponding to patch $\mathbf{p}_i$, $\mathbf{a}_i^j$ is the jth dimension of $\mathbf{a}_i$, and μ controls the sparseness. Using a gradient-based optimising method, $\mathbf{B}$ is first optimised only, keeping $\mathbf{a}_i$ for each patch fixed. After that, $\mathbf{B}$ is held and $\mathbf{a}$ is optimised. Using the optimised $\mathbf{B}$, a patch $\mathbf{p}$ is reconstructed by minimising the following equation:

$$F(\mathbf{a},\mathbf{B}) = \frac{1}{2} \| \mathbf{p} - \mathbf{a}\mathbf{B} \| + \mu \sum_{j}^{G} \mathbf{a}^j \quad . \tag{4.3}$$

Based on Eq. 4.3, the weight vector $\mathbf{a}$ for the patch $\mathbf{p}$ is obtained. In this work, each patch $\mathbf{p}$ is represented by a feature vector $\psi(\mathbf{p}) = \mathbf{a}$, where each dimension represents the weight of a base. Because the SC feature is sensitive to image rotations, each training image is rotated into 12 different positions (every $30°$ the image is rotated once) to consider a variety of rotations in SC feature extraction.

Besides the usage of feature description, SC is also useful in the task of image reconstruction and denoising. To reconstruct an image, there are two steps. The first step is to calculate the left term $F(\mathbf{a},\mathbf{B})$ in Eq. 4.3, which is the reconstructed vector of the image. The second step is to restore this obtained vector into the original layout of the matrix of the image. An example of the reconstructing ability of SC is shown in Figure 4.4(b).

To eliminate noise from an image, SC first uses a threshold in the image reconstructing process, where if the weight of a patch is lower than the threshold, it means this patch is less important and recognised as a background region in the image. Then, all the background regions are reconstructed using a unified colour, and the remaining "main body" is reconstructed in the normal way. An example of this image denoising

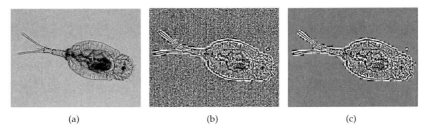

(a)   (b)   (c)

Figure 4.4: An example of SC for image reconstruction and denoising. (a) is an original image of *Rotifera*. (b) shows the reconstruction result of (a). (c) is the denoising result of (a).

approach using SC is shown in Figure 4.4(c). Compared to the normal reconstructed image in Figure 4.4(b), this result shows an obvious denoising performance.

4.2.2   Non-negative Sparse Coding

In contrast to the normal SC, NNSC is a conditional SC form, which has a non-negative condition to restrict the SC extracting process. To achieve NNSC, a brief way is to modify Eq. 4.3, where when any negative value in $\mathbf{a}$ or $\mathbf{B}$ appears, it is set to 0. Furthermore, using the optimised $\mathbf{B}$, a patch $\mathbf{p}$ is reconstructed by minimising the following NNSC equation:

$$F_{NN}(\mathbf{a}, \mathbf{B}) = \frac{1}{2} \| \mathbf{p} - \mathbf{aB} \| + \mu \sum_{j}^{G} a^{j} + \tau \| \mathbf{a} \|^{2} \quad ,$$

$$\text{subject to} \quad a^{j} \geqslant 0 \quad , \quad \mathbf{B} \geqslant 0 \quad , \quad \mu \geqslant 0 \quad \text{and} \quad 0 \geqslant \tau \geqslant -1 \quad . \tag{4.4}$$

Especially an asymmetric regularisation method introduced in [LRS10] is applied to enhance the sparsity of the final NNSC features in this dissertation. A quadratic term $\tau \| \mathbf{a} \|^{2}$ is used to further decrease the reconstruction cost. Based on Eq. 4.4 the weight vector $\mathbf{a}$ for the patch $\mathbf{p}$ is obtained. Similar to SC in this work, each patch $\mathbf{p}$ is represented by an NNSC feature vector $\psi_{NN}(\mathbf{p}) = \mathbf{a}$, where each dimension represents the weight of a base.

4.3   Weakly Supervised Learning

In contrast to SSL discussed in Section 3.5, weakly supervised learning (WSL) only needs labels (categorical information) of images, but does not need any additional annotations. Hence, WSL is useful for large scale image classification, where ground truth images (manually segmented images) are impossible to support. Furthermore, WSL can localise and classify interesting objects in an image synchronously, training a more advanced machine learning approach on the semantic level. In the following parts, two WSL frameworks are introduced. The first one is a basic region-based SVM (RBSVM), and the second one is an improved RBSVM.

4.3.1   Basic RBSVM

The basic RBSVM iterates to search the best subwindow for each training image, and to learn an SVM using newly found subwindows [Ngu+09]. Here, "the best subwindow" means a subwindow (or bounding box) which correctly and accurately contains the interesting object. Below, the best subwindow search and the RBSVM learning process is explained.

Best Subwindow Search

Assume that $\mathbf{d}$ is an image, $\mathbf{x}$ is a subwindow in $\mathbf{d}$, $\mathcal{X}$ denotes a set of subwindows in $\mathbf{d}$, and $\mathcal{S}$ is the set of all possible subwindows in $\mathbf{d}$ (i.e. $\mathcal{X} \subseteq \mathcal{S}$). Each $\mathbf{x}$ is defined by four position values $L_\mathbf{x}$, $T_\mathbf{x}$, $R_\mathbf{x}$, and $B_\mathbf{x}$ denoting left, top, right and bottom sides of $\mathbf{x}$, respectively. Similar to $\mathbf{x}$, $L_\mathbf{l}$, $T_\mathbf{l}$, $R_\mathbf{l}$, and $B_\mathbf{l}$ to denote the four sides of the largest possible subwindow, use $L_\mathbf{s}$, $T_\mathbf{s}$, $R_\mathbf{s}$, and $B_\mathbf{s}$ to denote the four sides of the smallest possible subwindow. For example, in Figure 4.5 all the subwindows between the largest and the smallest red ones construct $\mathcal{X}$. So, $\mathcal{X}$ contains all $\mathbf{x}$ restricted by $L_\mathbf{l} \leqslant L_\mathbf{x} \leqslant L_\mathbf{s}$, $T_\mathbf{s} \leqslant T_\mathbf{x} \leqslant T_\mathbf{l}$, $R_\mathbf{s} \leqslant R_\mathbf{x} \leqslant R_\mathbf{l}$, and $B_\mathbf{l} \leqslant B_\mathbf{x} \leqslant B_\mathbf{s}$. To search the best subwindow $\hat{\mathbf{x}}$ in $\mathcal{S}$, RBSVM uses a quality function f to evaluate the presence of an interesting object in $\mathbf{x}$. To perform a fast search for the best subwindow, an upper bound $\hat{f}$ that bounds the values of f over $\mathcal{X}$ is used. By this upper bound, many low quality subwindows can be effectively excluded and only the promising ones are kept. This upper bound $\hat{f}$ fulfills the following conditions:

$$
\begin{aligned}
&I) \quad \hat{f}(\mathcal{X}) \geqslant \max_{\mathbf{x} \in \mathcal{X}} f(\mathbf{x}) \quad, \\
&II) \quad \hat{f}(\mathcal{X}) = f(\mathbf{x}) \quad, \quad \text{if} \quad \mathbf{x} \quad \text{is the only element in} \quad \mathcal{X}.
\end{aligned}
\tag{4.5}
$$

$I)$ indicates that $\hat{f}$ represents the score which is larger than any score for all subwindows in $\mathcal{X}$. $II)$ means that if $\mathcal{X}$ only contains one subwindow $\mathbf{x}$, the upper bound score equals to the score of $\mathbf{x}$. When $\hat{f}$ has a higher score, $\mathcal{X}$ has a higher possibility of containing the interesting object. For example, Figure 4.5 shows the basic RBSVM using BoVW features [Ngu+09], where each image patch is assigned to a visual word

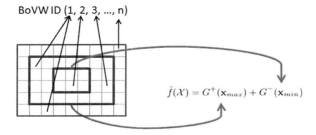

Figure 4.5: An example of the basic RBSVM using BoVW features.

and represented by its ID. For each subwindow $\mathbf{x}$, $\varphi(\mathbf{x})$ is used to represent its U-dimensional BoVW feature vector $\mathbf{c}_{\text{SIFT}}$ defined in Section 3.4, where each dimension indicates the frequency of a visual word in $\mathbf{x}$. The quality function f in the basic RBSVM using a linear SVM is defined as:

$$G(\mathbf{x}) = \mathbf{w}^{\mathsf{T}}\varphi(\mathbf{x}) + b = \sum_{z=1}^{Z} w(\mathbf{p}_z) + b \quad , \tag{4.6}$$

where $\mathbf{w} = (w_1, w_2, \ldots, w_G)$ represents the weight for each visual word, and b is the SVM bias. Here $\mathbf{p}_z(z = 1, 2, \ldots, Z)$ is the zth patch in subwindow $\mathbf{x}$, and $w(\mathbf{p}_z)$ is the weight of the visual word assigned to $\mathbf{p}_z$. In other words, the quality function $G(\mathbf{x})$ can be computed by summing weights of visual words assigned to patches in $\mathbf{x}$. Give a set of subwindows $\mathcal{X}$, the upper bound is defined as

$$\hat{f}(\mathcal{X}) = G^+(\mathbf{x}_{\text{max}}) + G^-(\mathbf{x}_{\text{min}}) + b \quad , \tag{4.7}$$

where $\mathbf{x}_{\text{max}}$ denotes the maximal subwindow in $\mathcal{X}$, $\mathbf{x}_{\text{min}}$ denotes the minimal subwindow in $\mathcal{X}$, $G^+(\mathbf{x}_{\text{max}})$ is the SVM score only using positive dimensions of $\mathbf{w}$, and $G^-(\mathbf{x}_{\text{min}})$ is the SVM score only using negative dimensions of $\mathbf{w}$. Because $G^+(\mathbf{x}_{\text{max}})$ produces the highest score of positive SVM weights, and $G^-(\mathbf{x}_{\text{min}})$ produces the highest score of negative SVM weights, the sum of them can be used as an upper bound of $\mathcal{X}$. Especially when $\mathcal{X}$ only contains one subwindow $\mathbf{x}$, $\mathbf{x}_{\text{max}} = \mathbf{x}_{\text{min}}$, so $\hat{f}(\mathcal{X}) = G(\mathbf{x})$. Using two integral images, the calculations of $\hat{f}(\mathcal{X})$ can be conducted in $O(1)$, where the first only contains patches related to $G^+(\mathbf{x}_{\text{max}})$, and the second only contains patches related to $G^-(\mathbf{x}_{\text{min}})$.

Basic RBSVM Training

To train a basic RBSVM classifier, the process is discussed as follows. Assume that $\mathcal{D}$ contains two subsets $\mathcal{D}^+$ and $\mathcal{D}^-$, which are sets of positive and negative images, respectively. Each image in $\mathcal{D}^+$ and $\mathcal{D}^-$ is labeled only with the presence or absence of an interesting object. A basic RBSVM is trained to localise and classify objects by solving the following constrained optimisation:

$$\begin{aligned}
\underset{\mathbf{w},b}{\text{minimise}} \quad & \zeta \sum_i (\alpha_i + \beta_i) + \frac{1}{2}\|\mathbf{w}\|^2 \quad , \quad \forall i \quad , \\
\text{subject to} \quad & \max_{\mathbf{x} \in \mathcal{X}_i^+} \{\mathbf{w}^\mathsf{T} \varphi(\mathbf{x}) + b\} \geqslant 1 - \alpha_i \quad , \quad \alpha_i \geqslant 0 \quad , \\
& \max_{\mathbf{x} \in \mathcal{X}_i^-} \{\mathbf{w}^\mathsf{T} \varphi(\mathbf{x}) + b\} \leqslant -1 + \beta_i \quad , \quad \beta_i \geqslant 0 \quad .
\end{aligned}$$
(4.8)

where, ζ is a penalty coefficient to penalise the mis-classification of each training example (training image). $\mathcal{X}_i^+$ is a subwindow set for the ith positive image, and $\mathcal{X}_i^-$ is a subwindow set for the ith negative image. α_i (or β_i) is the slack variable for the ith positive (or negative) image. The feature vector $\varphi(\mathbf{x})$ is computed for the searched best window $\mathbf{x}$. The first constraint means that each positive image should contain at least one subwindow which is associated with a positive value. The second constraint indicates that all subwindows in each negative image should be classified as negative values. Then, the SVM margin subject to these constraints must be maximised. To this end, the parameters ($\{\alpha_i\}, \{\beta_i\}, \mathbf{w}, b$) are optimised by a coordinate descent approach. For each of the four parameters, another three ones are held and a line search along the coordinate direction of it is performed. This parameter update is repeated until all parameters do not change. These optimised parameters are used to localise and classify an object in a test image $\mathbf{t} \in \mathcal{D}$. Basic RBSVM finds the subwindow $\hat{\mathbf{x}}$, which yields the maximal SVM score in $\mathbf{t}$:

$$\hat{\mathbf{x}} = \underset{\mathbf{x} \in \mathcal{X}}{\text{argmax}} \, \mathbf{w}^\mathsf{T} \varphi(\mathbf{t}) \quad .$$
(4.9)

If the value of $\mathbf{w}^\mathsf{T} \varphi(\hat{\mathbf{x}}) + b$ is positive, the basic RBSVM judges $\mathbf{t}$ as showing the object, otherwise $\mathbf{t}$ is judged as not showing it.

4.3.2   Improved RBSVM

Based on the basic RBSVM mentioned above, an improved RBSVM using SC features is proposed in this work. Primarily, to satisfy the upper bound condition, NNSC features are used.

Best Subwindow Search

In contrast to the basic RBSVM, the upper bound function $\hat{f}$ of the improved RBSVM is extended for NNSC features. As shown in Figure 4.5, the basic RBSVM method represents an image as one layer of visual words, but the improved RBSVM shown in Figure 4.6 using NNSC features applies different bases to decompose an image into multi-layers, where each layer represents weights of patches for a basis $\mathbf{b}_j$ in the basis set $\mathbf{B}$.

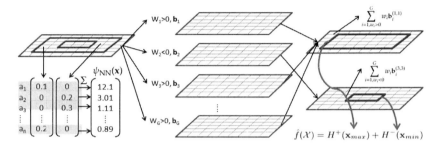

Figure 4.6: An example of the improved RBSVM using NNSC features.

Specifically, the zth patch $\mathbf{p}_z (z = 1,\ldots,Z)$ in a subwindow $\mathbf{x}$ is represented as a vector of non-negative sparse weights $\psi_{NN}(\mathbf{p}_z)$ as the definition in 4.2.2. For example in Figure 4.6, a patch is represented by a feature vector $(0.1, 0, 0, \ldots, 0.2)$. Furthermore, $\mathbf{x}$ is represented by the sum of all $\psi_{NN}(\mathbf{p}_z)$, which describes the accumulated weights of the basis in $\mathbf{x}$. Then, the quality function based on a linear SVM is reformulated as

$$
\begin{aligned}
H(\mathbf{x}) &= \mathbf{w}^{\mathsf{T}} \sum_{z=1}^{Z} \psi_{NN}(\mathbf{p}_z) + b \\
&= \sum_{j=1}^{G} w_j \sum_{z=1}^{Z} \psi_{NN}(\mathbf{p}_z)^j + b
\end{aligned}
\tag{4.10}
$$

where $\psi_{NN}(\mathbf{p}_z)^j$ represents the weight of $\mathbf{p}_z$ for the jth basis $\mathbf{b}_j$. In Eq.4.10, the summation of weights of patches ($\sum_{z=1}^{Z} \psi_{NN}(\mathbf{p}_z)^j$) is multiplied by the SVM weight w_j. Thus, the upper bound for $\mathcal{X}$ is defined in the following way:

$$
\hat{f}(\mathcal{X}) = H^+(\mathbf{x}_{\max}) + H^-(\mathbf{x}_{\min}) + b \quad,
$$

$$
H^+(\mathbf{x}_{\max}) = \sum_{j=1, w_j \geq 0}^{G} w_j \psi_{NN}(\mathbf{x})^j \quad,
$$

$$
H^-(\mathbf{x}_{\min}) = \sum_{j=1, w_j < 0}^{G} w_j \psi_{NN}(\mathbf{x})^j \quad,
\tag{4.11}
$$

where $\psi_{NN}(\mathbf{x})^j = \sum_{z=1}^{Z} \psi_{NN}(\mathbf{p}_z)^j$, which is illustrated by the horizontal summation in Figure 4.6. Because NNSC features are used in this work, $\psi_{NN}(\mathbf{x})^j$ always take non-negative values. Thus, the layers with the same signs of w_j are grouped into a single layer. On the one hand, $H^+(\mathbf{x}_{max})$ is computed for the maximal subwindow $\mathbf{x}_{max}$ in the layer built based on positive w_j. On the other hand, $H^-(\mathbf{x}_{min})$ is computed for the minimal subwindow $\mathbf{x}_{min}$ using the layer built based on negative w_j. Similar to the basic RBSVM, $\hat{f}(\mathcal{X})$ in Eq. 4.11 represents the score which is larger than any score for all subwindows in $\mathcal{X}$. In addition, if $\mathbf{x}_{max} = \mathbf{x}_{min}$, $\hat{f}(\mathcal{X})$ becomes $H(\mathbf{x})$.

Improved RBSVM Training

Similar to the training process of the basic RBSVM, assume that $\mathcal{D}$ contains two subsets $\mathcal{D}^+$ and $\mathcal{D}^-$, which are sets of positive and negative images, separately. Each image in $\mathcal{D}^+$ and $\mathcal{D}^-$ is labeled only with the presence or absence of an interesting object. An improved RBSVM is trained to localise and classify objects by optimising the following constrained equation:

$$\begin{aligned}
\underset{\mathbf{w},b}{\text{minimise}} \quad & \zeta \sum_i (\alpha_i + \beta_i) + \frac{1}{2} \| \mathbf{w} \|^2 \quad , \quad \forall i \quad , \\
\text{subject to} \quad & \max_{\mathbf{x} \in \mathcal{X}_i^+} \{\mathbf{w}^\mathsf{T} \psi_{NN}(\mathbf{x}) + b\} \geqslant 1 - \alpha_i \quad , \quad \alpha_i \geqslant 0 \quad , \\
& \max_{\mathbf{x} \in \mathcal{X}_i^-} \{\mathbf{w}^\mathsf{T} \psi_{NN}(\mathbf{x}) + b\} \leqslant -1 + \beta_i \quad , \quad \beta_i \geqslant 0 \quad .
\end{aligned} \tag{4.12}$$

where ζ is a penalty coefficient to penalise the mis-classification of each training example (training image). $\mathcal{X}_i^+$ is a subwindow set for the ith positive image, and $\mathcal{X}_i^-$ is a subwindow set for the ith negative image. α_i (or β_i) is the slack variable for the ith positive (or negative) image. The feature vector $\psi_{NN}(\mathbf{x})$ is computed for the searched best window $\mathbf{x}$. The first constraint means that each positive image should contain at least one subwindow which is associated with a positive value. The second constraint indicates that all subwindows in each negative image should be classified as negative values. Then, the SVM margin subject to these constraints needs to be maximised. To this end, the parameters $(\{\alpha_i\}, \{\beta_i\}, \mathbf{w}, b)$ are optimised by a coordinate descent approach as the one in Section 4.3.1. These optimised parameters are used to localise and classify an object in a test image $\mathbf{t} \in \mathcal{D}$. Improved RBSVM finds the subwindow $\hat{\mathbf{x}}$, which yields the maximal SVM score in $\mathbf{t}$:

$$\hat{\mathbf{x}} = \underset{\mathbf{x} \in \mathcal{X}}{\text{argmax}} \; \mathbf{w}^\mathsf{T} \psi_{NN}(\mathbf{t}) \quad . \tag{4.13}$$

If the value of $\mathbf{w}^\mathsf{T} \psi_{NN}(\hat{\mathbf{x}}) + b$ is positive, the improved RBSVM judges $\mathbf{t}$ as showing the object, otherwise $\mathbf{t}$ is judged as not showing it.

Furthermore, the late fusion approach that is introduced in Section 3.5.3 can also be applied here to enhance the overview performance of classification.

4.4 Summary

In this chapter, the full-automatic microscopic image classification system using WSL is proposed. Firstly, an overview of this system is introduced in Section 4.1. Then, SC features are presented in Section 4.2. Finally, WSL classifier is proposed in Section 4.3, where it can classify and localize an object jointly.

Chapter 5

Microscopic Image Clustering Using Unsupervised Learning

In this chapter, a microscopic image clustering system is proposed using an unsupervised learning approach. An overview of this system is first introduced in Section 5.1. Then, a full-automatic image segmentation method is proposed in Section 5.2. Thirdly, several selected feature extraction methods are introduced in Section 5.3. Next, an unsupervised learning algorithm is presented in Section 5.4. Lastly, a brief conclusion is given at the end of this chapter in Section 5.5. Especially microscopic images of stem cells in migration are used in this chapter for a clear explanation of the proposed algorithms.

5.1   Overview of the System

In contrast to the classification tasks using supervised learning methods (see Chapter 3 and 4), unsupervised learning is usually applied for clustering tasks to help researchers analyse and summarize characteristics and properties of the data, where original data has no specific information of categories. To this end, a microscopic image clustering system is built to group microscopic images that have similar features into the same classes (or clusters). In this system, three problems are addressed by CBMIA approaches. The first is image segmentation, where a double-stage segmentation approach (see Section 5.2) is developed. The second is shape description, which is solved by various global shape features in Section 5.3. The third problem is clustering, which is addressed by a k-means clustering approach (see Section 5.4).

For an intuitive understanding, Figure 5.1 shows the working flow of this system. In (a), original microscopic images of stem cells are first used as input data. Then, the double-stage segmentation method is used to localise the shape of each stem cell in (b). Thirdly, different global features are used to describe the shapes of these segmented cells in (c). Fourthly, (d) shows that the most effective features used above are selected and fused to improve the clustering performance. Finally, in (e), k-means clustering is applied to implement an unsupervised learning approach of the stem cells using the fused features.

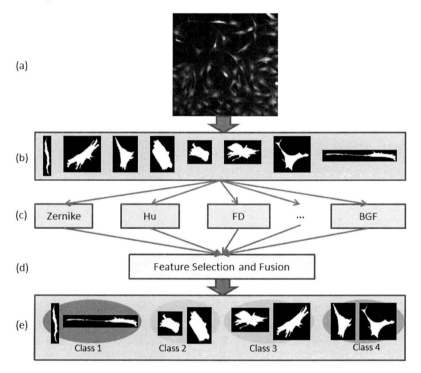

Figure 5.1: Working flow of the proposed stem cell clustering system. (a) is an original stem cell microscopic image. (b) shows the segmented stem cells. (c) is the feature extraction approach. (d) shows feature selection and fusion. (e) is the final classification result, where cells with similar shapes are classified into the same category.

5.2 Full-automatic Image Segmentation

5.2.1 Motivation

In some microscopic images, objects contain very tiny parts that include important detailed information, e.g. the filamentous cytoplasm of stem cells in Figure 5.2(a) is a significant component in the cell structure. However, normal segmentation approaches mainly contain the main parts of an object and ignore the details, leading to an over-segmentation problem. To overcome this problem, a full-automatic image segmentation approach, namely 'double-stage segmentation', is developed using Sobel edge detection.

5.2.2 Techniques

This segmentation method contains two main stages: The first stage is defined as follows

$$
\text{Region} = \begin{cases} \text{Main Part} \to \text{Keep} & \text{if} \quad \text{Region's Size} \geqslant \text{High Threshold} \\ \text{Others} \to \text{Remove} & \text{if} \quad \text{Region's Size} < \text{High Threshold} \end{cases}, \quad (5.1)
$$

where the algorithm only segments the main parts of the object in this stage. When the size of a segmented region (Region's Size) is larger than a given threshold (High Threshold), this region is kept as a main part, otherwise it is removed from the image.

In contrast to the first stage, the second stage only segments the details of the object, and it is defined as

$$
\text{Region} = \begin{cases} \text{Detail} \to \text{Keep} & \text{if} \quad \text{Region's Size} \leqslant \text{Low Threshold} \\ \text{Others} \to \text{Remove} & \text{if} \quad \text{Region's Size} > \text{Low Threshold} \end{cases}, \quad (5.2)
$$

where, if the size of a segmented region (Region's Size) is smaller than a given threshold (Low Threshold), this region is kept as a detailed part (Detail), otherwise it is removed from the image. Note that: High Threshold > Low Threshold, and they are both obtained based on previous experiments.

After the aforementioned two stages, their segmentation results are combined together. In this combination, the matrixes of two segmented images are added first, then, the High Threshold in Eq. 5.1 is applied again to the new image, lastly, the remaining parts are converted to a binary image which is the final segmentation result. An example of this segmentation process is shown in Figure 5.2.

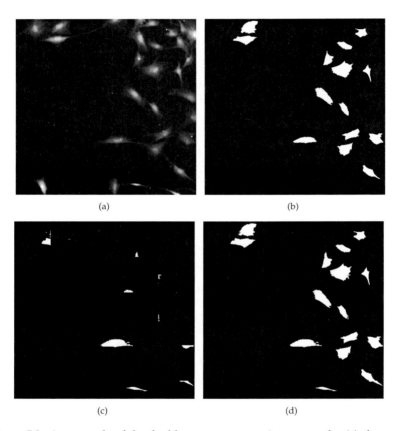

(a)   (b)

(c)   (d)

Figure 5.2: An example of the double-stage segmentation approach. (a) shows an original microscopic image of stem cells. (b) is the segmentation result of the first stage. (c) shows the segmentation result of the second stage. (d) is the final segmentation result combined by (b) and (c).

5.3 Global Shape Features

In this section, the selected global shape feature extraction techniques are illuminated, including edge histogram descriptor (EHD), higher-level geometrical feature (HGF), and shape context feature (SCF). For all the mentioned features, the input data is a binary image (the segmentation result of Section 5.2), and the output data is a feature vector. Especially because images of environmental microorganisms (EMs) are already explicitly classified data, they are used as examples to explain the discriminative characteristics of these features, where the region of an EM (in white) is distinguished from the background (in black).

5.3.1 Edge Histogram Descriptor

EHD is a contour-based shape feature, which is powerful to describe objects that contain linear edges [FG11]. In EHD, the edges in an image are first grouped into five categories by their approximate directions, including vertical ($0°$), $45°$, horizontal ($90°$), $135°$, and uncertain orientations. An example of the five types of edges is shown in Figure 5.3.

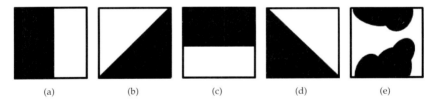

(a)   (b)   (c)   (d)   (e)

Figure 5.3: Five types of edges. (a) Vertical ($0°$) edge. (b) $45°$ edges. (c) Horizontal ($90°$) edge. (d) $135°$ edge. (e) Uncertain edge.

Then, the amount of edges of each type is calculated to build a histogram, namely the "edge histogram". In this histogram, the value of each bin represents the statistical characteristic of its corresponding edge type. In this work, the order of bins in EHD is ranked following ($0°, 45°, 90°, 135°$, uncertain) edges. In Figure 5.4, an example of EHD is shown, where the histograms of two EMs are compared to show the discriminative power of EHD. Because the *Actinophrys* in Figure 5.4(a) has a lot of filamentous pseudopodium, it contains more edges than the *Aspidisca* in Figure 5.4(b), leading to totally different histograms in Figure 5.4(c).

Although EHD is an effective shape feature, it is sensitive to object rotation. When the same objects state in different directions, their EHDs possibly obtain entirely

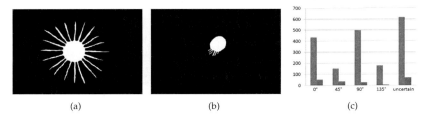

Figure 5.4: An example of EHD. (a) An image of *Actinophrys*. (b) An image of *Aspidisca*. (c) EHDs of (a) and (b). The blue bins represent the EHD of (a), and the red bins show the EHD of (b). The horizontal axis shows the five types of edges. The vertical axis shows the amount of edges.

different distributions. For example, the EHDs of two objects in Figure 5.5(a) and Figure 5.5(b) are compared in Figure 5.5(c), where they have the same shapes, but state in two directions. To solve this rotation problem of EHD, a simple ranking strategy is applied, where the five bins are resorted by an ascending order and structure a new histogram. This modified EHD eliminates the sensitivity of the original EHD, so it is used in this work instead of the original one. An example of this new EHD is shown in Figure 5.5(d), where the rotation problem that occurs in Figure 5.5(c) is resolved. Finally, a five-dimensional feature vector is obtained based on this histogram: $c_{EHD} = (EHD_1, EHD_2, \ldots, EHD_5)^T$, where $EHD_{1,2,\ldots,5}$ is the value of each bin, respectively.

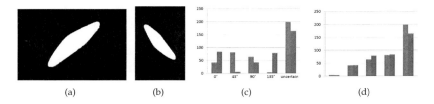

Figure 5.5: An example of EHD in different directions. (a) An image of *Paramecium* (deviation to the right). (b) An image of *Paramecium* (deviation to the left). (c) EHDs of (a) and (b), where the horizontal axis shows the five types of edges. (d) Modified EHDs of (a) and (b), where the horizontal axis shows edges in an ascending order. The blue bins represent the edges of (a), and the red bins show the edges of (b). The vertical axis displays the amount of edges.

5.3.2 Higher-level Geometrical Feature

In this section, two effective HGFs are discussed. The first HGF is *Hu Moment* (Hu) [Hu62], and the second is *Zernike Moment* (Zernike) [TSS11].

Hu Moment

Hu moment is also known as the 'invariant moment', which is invariant to describe an object in multiple deformations, e.g. translation, changes in scale, and also rotation. In a given image with the size of $M \times N$ pixels, the pixel value at the position (x, y) is $f(x, y)$, then, the $(p + q)$ order raw moment (or geometrical moment) m_{pq} and central moment μ_{pq} are first defined as follows

$$m_{pq} = \sum_{x=1}^{M} \sum_{y=1}^{N} x^p y^q f(x, y) \quad ,$$

$$\mu_{pq} = \sum_{x=1}^{M} \sum_{y=1}^{N} (x - \overline{x})^p (y - \overline{y})^q f(x, y) \quad , \quad p, q = 1, 2, \ldots \quad ,$$

(5.3)

where $\overline{x} = m_{10}/m_{00}$ and $\overline{y} = m_{01}/m_{00}$. $(\overline{x}, \overline{y})$ is the centre of gravity of the image, where because binary images are used as the input data, it is also the centre of the object in the image. Furthermore, to make the obtained μ_{pq} have the invariant property, a normalising process is used to eliminate the original information of the scales of the object. This normalisation is given in the following

$$\eta_{pq} = \frac{\mu_{pq}}{\mu_{00}^{[1+(p+q)/2]}} \quad , \quad p + q = 2, 3, \ldots \quad .$$

(5.4)

Based on the normalised central moment η_{pq} mentioned above, seven Hu moments are defined as follows

$$\begin{aligned}
\mathrm{Hu}_1 &= \eta_{20} + \eta_{02} \\
\mathrm{Hu}_2 &= (\eta_{20} - \eta_{02})^2 + 4\eta_{11}^2 \\
\mathrm{Hu}_3 &= (\eta_{30} - 3\eta_{12})^2 + (3\eta_{21} - \eta_{03})^2 \\
\mathrm{Hu}_4 &= (\eta_{30} + \eta_{12})^2 + (\eta_{21} + \eta_{03})^2 \\
\mathrm{Hu}_5 &= (\eta_{30} - 3\eta_{12})(\eta_{30} + \eta_{12})[(\eta_{30} + \eta_{12})^2 - 3(\eta_{21} + \eta_{03})^2] \\
&\quad + (3\eta_{21} - \eta_{03})(\eta_{21} + \eta_{03})[3(\eta_{30} + \eta_{12})^2 - (\eta_{21} + \eta_{03})^2] \\
\mathrm{Hu}_6 &= (\eta_{20} - \eta_{02})[(\eta_{30} + \eta_{12})^2 - (\eta_{21} + \eta_{03})^2] + 4\eta_{11}(\eta_{30} + \eta_{12})(\eta_{21} + \eta_{03}) \\
\mathrm{Hu}_7 &= (3\eta_{21} - \eta_{03})(\eta_{30} + \eta_{12})[(\eta_{30} + \eta_{12})^2 - 3(\eta_{21} + \eta_{03})^2] \\
&\quad - (\eta_{30} - 3\eta_{12})(\eta_{21} + \eta_{03})[3(\eta_{30} + \eta_{12})^2 - (\eta_{21} + \eta_{03})^2] \quad .
\end{aligned}$$

(5.5)

These seven Hu moments can be combined into a seven-dimensional feature vector
$c_{Hu} = (Hu_1, Hu_2, \ldots, Hu_7)^T$. An example of Hu moments is shown in Figure 5.6, where
a visible histogram representation is used to demonstrate the usage of this feature.
By this example, it is easy to find that Hu moments have similar values for the object
with different deformations.

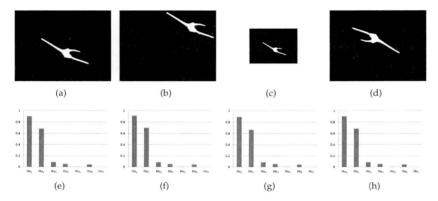

Figure 5.6: An example of Hu moments. (a) is the original image of *Ceratium*. (b),
(c), and (d) are the deformed results of (a) with translation, 0.5 time scaled and 180°
rotation, respectively. (e), (f), (g), and (h) are the visible histograms of Hu moments
of (a), (b), (c), and (d), separately. The horizontal axis in each histogram shows seven
Hu moments, and the vertical axis shows their normalised values in [0, 1].

Zernike Moment

Zernike moment is the projection of the shape function of an object on orthogonal
polynomial [KH90; Sak+13], which has an orthogonality and is robust to the image
rotation problem [TSS11]. First, assume that $f(x, y)$ is the pixel value of position (x, y)
in a given image. Then, the orthogonal polynomial of Zernike moment in unit circle
is defined as

$$V_{nm}(x, y) = V_{nm}(\rho, \theta) = R_{nm} \exp(jm\theta) \quad , \tag{5.6}$$

where $n = 0, 1, 2, \ldots$, $m = \ldots, -2, -1, 1, 2, \ldots$, and $n - |m| =$ even, $|m| \leq n$, ρ is the vector
from origin of coordinate to point (x, y), θ is the angle between ρ and x axis, $R_{nm}(\rho)$ is
the radial polynomial. Then, an n order Zernike moment of the image is defined as

follows [1]

$$\text{Zernik}_{nm} = \frac{n+1}{\pi} \sum_x \sum_y f(x,y) V^*_{nm}(\rho,\theta) \quad , \quad \text{where} \quad x^2 + y^2 \leq 1 \quad . \tag{5.7}$$

Where $*$ denotes the complex conjugate. Because Zernik_{nm} has a high order invariant property, it is robust to the image rotation problem. It is worth mentioning that in this work, $n = 4$ and $m = 2$ are set based on pretests, leading to a single value feature vector $\mathbf{c}_{\text{Zernik}} = (\text{Zernik}_{42})^T$.

5.3.3 Shape Signature Histogram

SSH supports another simple strategy to solve the rotation problem of shape signature (SS) [ZL04]. Based on SS mentioned in Section 3.3.3, all the centroid distances d_t are assigned to different groups by their lengths. Then, the number of d_t in each group is calculated for building up a histogram, where each bin of the histogram represents the number of d_t in a group. Because this histogram is only the statistics of length of SS, it eliminates the influence of image rotation. In this work, $K = 100$ and 180 equidistant bins are used based on pretests, finally leading to a 180-dimensional feature vector $\mathbf{c}_{\text{SSH}} = (\text{SSH}_1, \text{SSH}_2, \ldots, \text{SSH}_{180})^T$, where $\text{SSH}_{1,2,\ldots,180}$ are the numbers of the 180 bins, respectively.

5.3.4 Shape Context Feature

SCF is a global shape feature, which is usually used for tasks of shape matching [BMP02]. First, K equidistant points on the contour of an object are sampled. Second, the vectors from each point to all remaining points are calculated. Finally, all these vectors constitute the SCF for this point, which represent the spatial relations between this point and the whole object. Because the basic SCF is sensitive to image rotation, the absolute values of these SCFs are used to remove the directivity of the vectors. An example of SCFs is shown in Figure 5.7.

In this work, $K = 200$ is used based on pretests, so for each point a 199-dimensional SCF feature vector is obtained. To use these obtained vectors to represent the object, a max-pooling approach is applied, where the biggest value of each dimension of all the vectors is selected and combines a final 199-dimensional SCF feature vector $\mathbf{c}_{\text{SCF}} = (\text{SCF}_1, \text{SCF}_2, \ldots, \text{SCF}_{199})^T$, where $\text{SCF}_{1,2,\ldots,199}$ is the result of max-pooling of each dimensionality, separately.

---

[1]For more mathematical information, please refer to Jamie Shutler's note, 15-08-2002: http://homepages.inf.ed.ac.uk/rbf/CVonline/ LOCAL-COPIES/SHUTLER3/node11.html.

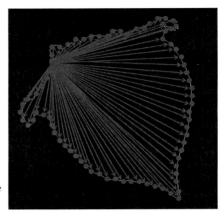

Figure 5.7: An example of SCF on an image of *Synchaeta*.

Besides these four global shape features introduced above, other features that are proposed in Section 3.3 can also be applied in this clustering task, including IEHD, SS, FD, BGF and ISH.

5.4 Unsupervised Learning

In contrast to the supervised learning methods, unsupervised learning does not need any annotations of the original data. Hence, it is usually used as an automatic classification approach of data mining and data analysis. Especially because the k-means clustering method can solve a large scale data case very robustly, it is selected in this work.

5.4.1 *k*-means Clustering

Assume a set of feature vectors is first given as $\mathcal{U} = \{\mathbf{u}_1, \mathbf{u}_2, \ldots, \mathbf{u}_i, \ldots, \mathbf{u}_{|\mathcal{U}|}\}$, where $|\mathcal{U}|$ is the number of feature vectors in $\mathcal{U}$ and $\mathbf{u}_i$ is the ith feature vector. Then, k-means clustering aims to group all these $|\mathcal{U}|$ data points (the feature vectors) into k classes (or clusters). To this end, an iterative algorithm is applied to minimise the sum of distances between the data points and k clustering centres, where a batch update strategy is employed in three steps. In the first step, all the points are randomly assigned to k clusters as an initialisation. In the second step, Euclidean distance is used to measure the dissimilarity between each of $\mathbf{u}_i$ and the centres $\mathbf{c}_{j=1,2,\ldots,k}$, where each $\mathbf{c}_j$ relates to the cluster c_j. Finally, the centre of each cluster is iteratively calculated to reassign these points to their nearest centres to determine a new cluster, where the iteration finishes until the new cluster centre does not have any massive change. This k-means clustering process is mathematically defined as follows:

$$\underset{\mathbb{C}}{\operatorname{argmax}} \sum_{j=1}^{k} \sum_{\mathbf{u}_i \in c_j} \|\mathbf{u}_i - \mathbf{c}_j\|^2 \quad , \tag{5.8}$$

where $\mathbb{C} = \{\mathbf{c}_1, \mathbf{c}_2, \ldots, \mathbf{c}_k\}$ is the set of all k clusters. A visible example of the k-means clustering is shown in Figure 5.8, where the points are finally grouped into their nearest clusters, respectively.

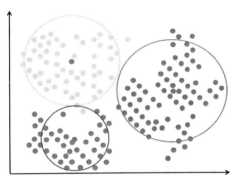

Figure 5.8: An example of k-means clustering. The red, orange and blue points show three classes of data. The purple points are the clustering centres of these three classes, respectively. The red, orange and blue circles represent the ranges of Euclidean distances of these three classes.

5.4.2 Clustering Evaluation

To evaluate the clustering quality, there are two usual methods. The first is the *Mean Variance Measurement*, which is a quantitative evaluation method. The second is the *Silhouette Plot* method, which is a visual evaluation approach.

Mean Variance Evaluation

Mean variance function is usually used as a standard measure for quantitatively evaluating the clustering results, calculating the mean variance of the distances between each data point and its clustering centre. When the clustering result is good, this value approaches 1. Otherwise, if the clustering result is bad, this value goes to 0. The definition of mean variance is as follows

$$J = 1 - \sqrt{\sum_{i=1}^{|\mathcal{U}|} \sum_{j=1}^{k} \frac{(\mathbf{u}_{ij} - \mathbf{c}_j)^2}{|\mathcal{U}| - 1}} \tag{5.9}$$

where $\mathbf{u}_{ij}$ means that $\mathbf{u}_i$ is classified into the jth cluster. If the clustered points in each cluster are more concentrated (convergent) to their centres respectively, J obtains a higher value. Thus, it can be used to evaluate the clustering result.

Silhouette Plot Evaluation

In order to have a visualized impression of a clustering result, a silhouette plot is used to illustrate the degree of convergence of all the data points. The silhouette plot shows a measurement of how close each point in a cluster is to points in its neighbouring clusters. The range of this measurement is $[-1, 1]$, where $[-1, 0)$ indicates points that are probably grouped into a wrong cluster, 0 indicates points that are not in between of one and another cluster, and $(0, 1]$ indicates points that are far away from neighbouring clusters. An example of the silhouette plot for clustering evaluation is shown in Figure 5.9.

Figure 5.9: An example of the silhouette plot for clustering evaluation. The horizontal axis shows the value of the silhouette plot's measurement. The vertical axis shows the clustering numbers of points in each cluster.

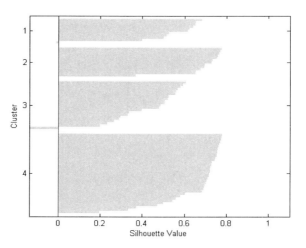

From Figure 5.9, people can gather that most points in the fourth cluster have a large value that is higher than 0.6, indicating that this cluster is far away from its neighbouring clusters. In contrast, the first and second clusters have a lot of points with low values, and the third one contains some points with negative values, indicating those three clusters are not very concentrated. Furthermore, the silhouette value supports a quantitative approach to evaluate the clustering result, where the average value is used to measure the overall performance of the clustering quality. If this average value is high, the overall performance is good. Otherwise, the overall performance is bad [2].

5.5   Summary

In this chapter, the proposed microscopic image clustering system is built using an unsupervised learning approach. The overview of this system is first introduced in Section 5.1. Then, a full-automatic image segmentation method is developed in Section 5.2 based on Sobel edge detection. Thirdly, four global shape features, EHD, HGF, SSH and SCF are stated in Section 5.3 to describe the shapes of objects. Finally, a k-means clustering algorithm is selected to solve the unsupervised learning problem.

---

[2]Silhouette plot of k-Means Clustering: http://de.mathworks.com/help/stats/k-means-clustering.html#brah7fp-1

Chapter 6

Object Tracking Using Interactive Learning

In this chapter, a novel object tracking approach is proposed to solve the problems mentioned in Section 2.1.6. To this end, the fundamental conception is first stated in Section 6.1. Then, techniques of preliminaries are introduced in Section 6.2. Thirdly, an effective interactive learning based object tracking framework is proposed in Section 6.3. Finally, a conclusion closes this chapter in 6.4. Especially, honey bees' micro-alike shots (videos) are used in this chapter as examples to explain the proposed method [1].

6.1 Fundamental Conception

The movements of body parts of harnessed insects, such as antennae of mouthparts, provide information about internal states [Sau+03], sensory processing [Erb+93] and learning [Reh87; SAT91; Cha+06; Gd10]. Although there is some research reported the field of animal tracking, estimating the center of body mass (position) is much simpler than detecting the detailed body posture and position of appendages (pose) [PE+14]. To the best of my knowledge, this work is the first research on the topic of tracking multiple insect body parts that are of different types. Insect posture is estimated as the tip of each body part (e.g. a bee's antennae or tongue as shown in Figure 6.1).

Although the application scenario of this tracking framework addresses a particular task, the challenges to be addressed, however, characterize a generic tracking problem resulting from: 1) a varying number of targets, 2) incoherent motion, 3) occlusion and merges, 4) all targets have a dark appearance, similar shape and no

---

[1]Note: because videos are investigated in this chapter, which are different from the 2-D images discussed in the former chapters, the mathematical symbols are redefined.

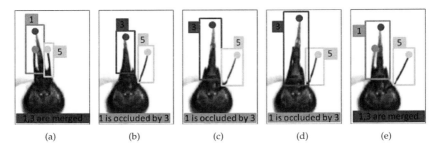

Figure 6.1: Sample frames of (a, e) merge or (b, c, d) occlusion. Merged targets are difficult to be differentiated at bounding box level, thus it is proposed to estimate the position of the tip of each target, which is denoted as a solid circle in the corresponding colour.

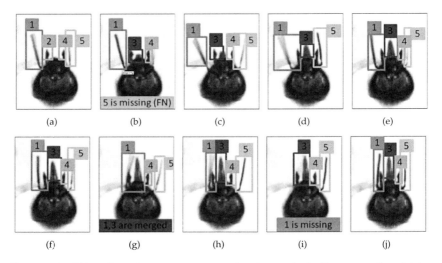

Figure 6.2: Object detections at ten consecutive frames including merged and false negative bounding boxes. The identification of each bounding box, shown in a different colour, is a challenging task. The label for each body part is denoted as 1: Left antenna; 2: Left mandible; 3: Proboscis; 4: Right mandible; 5: Right antenna.

texture and 5) long tracking gaps. Most tracking frameworks assume coherent motion, i.e. all the elementary targets move with similar average velocity over extended periods of time. However, this assumption does not hold here. A pictorial illustration is shown in Figure 6.2, where a set of object detections as unordered bounding boxes are produced by a standard moving object detector. Different colours are used here to denote the expected label for better visualization. It can be seen that a merged (see Figure 6.2(g)) or false negative (FN) bounding box (see Figure 6.2(b) and 6.2(i)) produces a tracking gap, which makes it unsuitable for frame-by-frame tracking approaches such as particle filter based algorithms. As the mandibles (i.e. label 2 and 4) do not provide much information for biologists, they are not tracked in the case where they are merged or occluded.

The different occlusion and merge conditions are illustrated in Figure 6.1. These issues were already addressed in the previous work [She+14], but the targets are difficult to differentiate at bounding box level under merge conditions (see Figure 6.1(a) and 6.1(e)). This work denotes *occlusion* as the cases where target a is occluded by target b, and *merge* where targets a and b are merged at the same bounding box. For occlusion conditions, estimating the position of an occluded target a if it is not visible makes little sense, though maintaining its identity when it appears again is challenging. For merge conditions, a new algorithm is proposed to differentiate targets at pixel precision by estimating the tip of each target (shown as the small solid circle in Figure 6.1(a) and 6.1(e).

The tracking problem of this work is formulated as follows. The inputs to the proposed tracking framework are a set of detection responses at bounding box level, thus only providing a rough estimation of the targets' position. The detection responses are denoted by $\mathbf{Z}_{1:N} = \{\mathbf{z}_{i,t} | 1 \leq i \leq n_t, 1 \leq t \leq N\}$, where n_t is the number of detection responses at time t. The objective is to estimate the trajectories of the tips of n targets. In the case of a honey bee, $n = 5$, 1: right antenna; 2: right mandible; 3: proboscis; 4: left mandible; 5: left antenna. The trajectories are denoted as $\mathbf{T} = \{T^i_{t_{i1},t_{i2}} | 1 \leq i \leq n\}$, where $T^i_{t_{i1},t_{i2}}$ is the track of the ith target existing from time t_{i1} to t_{i2}.

In this work, an interactive learning framework for insect tracking integrating a frame query approach is proposed, instead of the traditional track-and-then-rectification scheme. As shown in Figure 6.3, the overall framework includes six stages: (1) moving object detection, (2) feature extraction, (3) classification of moving objects, (4) constrained frame-to-frame linking, (5) *Key Frame* (KF) estimation and annotation query and (6) track linking through merge conditions. The yellow blocks highlight the interactive part, while the blue blocks indicate the automated computation part. The tracking problem is addressed by fulfilling two sub-tasks. The first sub-task is to assign a label $y_{i,t}$ to the corresponding bounding box $\mathbf{z}_{i,t}$, and construct tracks at bounding box level $\mathbf{Y}_{1:N} = \{y_{i,t} | 1 \leq i \leq n_t, 1 \leq t \leq N\}$. Given the input $\mathbf{z}_{i,t}$, a feature vector $\mathbf{f}_{i,t}$ is extracted to represent the information about its position, motion and shape. The initial label $y_{i,t}$ is estimated by classification (Section 6.3.1) and constraint frame-to-frame linking (Section 6.3.2). This framework queries users to rectify the incorrect labels only for certain frames (i.e. $Y_s | s \in \Phi$, where Φ is the set of KFs), which are estimated in Section 6.3.3, and the framework takes them as prior information to compute the labels of bounding boxes on the other frames. The tracks are iteratively refined until a user query is no longer required. As a result, reliable tracks $\mathbf{Y}_{1:N}$ are constructed, which is indicated with a pink shaded ellipse in Figure 6.3(a). The second sub-task is to find the position of the tip (i.e. the endpoint, shown as coloured solid circles in Figure 6.1) of each target $\mathbf{x}_t^i$ and construct complete tracks $\mathbf{T} = \{T_{t_{i1},t_{i2}}^i | 1 \le i \le n\}$ through merge or occlusion conditions, which are indicated as solid coloured lines in Figure 6.3(b). An algorithm is proposed in Section 6.3.4 to link the gaps between the tracks to automatically compute the final trajectories $\mathbf{T}$.

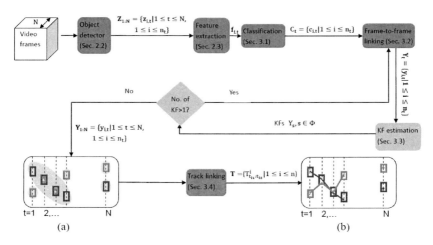

Figure 6.3: The flowchart of the overall tracking framework: the yellow blocks highlight the interactive part, while the blue blocks denote the automated computation part.

Furthermore, to overcome the bottleneck of the automatic tracking performance by introducing user input, some interactive algorithms have been reported [Yao+12; Yue+09; BF06; VR11]. But some of them either require users to view the whole video [Yue+09; Yao+12], or not to focus on frame query techniques [BF06]. The most conceptually similar work to the method introduced in this dissertation is proposed in [VR11], which extends the tracker in [VRP10] by estimating more KFs for user annotation to improve the tracking accuracy. However, since the KF estimation scheme in [VR11] punishes significant label change, it is not applicable to the micro-alike shot (or macroshot) analysing tasks, where different objects could be detected in turns at the same position as the example shown in Figure 6.2.

6.2 Preliminaries

When controlled stimulus conditions are needed, insects are often restrained and their behavior is monitored as movements of body parts such as their antenna or mouth-parts. The proboscis is the mouthpart of the insect, and hungry bees extend their proboscises reflexively when stimulated with food or with a previously conditioned odorant as shown in Figure 6.4.

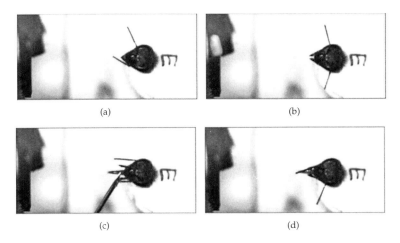

(a)   (b)

(c)   (d)

Figure 6.4: Associative odor-reward learning paradigm in honey bees. A bee that has learnt the association between an odorant and a food-reward extends its proboscis when stimulated with the learnt odorant: (a) Before odorant stimulation. (b) Odorant released indicated by the LED. (c) Sugar rewarding. (d) During odorant stimulation.

6.2.1   Object Detection and Preprocessing

As the interest of this work focuses on tracking the antennae and mouthparts of a bee when they are moving, it is preferred to detect the moving part rather than segment the body part on a single frame basis. Thus, *Gaussian Mixture Model* (GMM) background modeling [KB01] is used. A more advanced background subtraction method based on a dynamic background model [PAV15] may reduce false detections, but a standard moving object detector is used here as this work focuses on the tracking part. The object detector generates an unordered set of bounding boxes including false positives (e.g. shadows, reflection and the insect's legs), false negatives (e.g. motion blurred antennae), missing objects (e.g. the antenna above the insect's head, or the proboscis not extended), merged detections (one bounding box including two or more objects) and occluded detections, which make the following tracking task difficult. Therefore, preprocessing operations include shadow removal [KB01], exclusion of undesired objects by incorporating position information, and the segmentation of merged measurements.

These preprocessing operations greatly reduce the undesired detection measurement, but some false, missing, merged measurements may still remain. Thus, a subsequent tracking algorithm is required to tackle this problem.

6.2.2   Anatomical Model of Insect Body Parts

Modeling the anatomy of an insect's head is important for accurate tracking, due to the physical limitations of the moving objects' relative positions. The positions of insect body parts (e.g. antennae and mouthparts) are ordered in a certain sequence, which is rather similar among various insects. Figure 6.5 shows the image of an ant's head. These body parts are symmetric, thus, they could be classified according to their types, and then further identified (tracked) exploiting temporal correlation between neighboring frames. The proposed framework incorporates an anatomical model of an insects' head as a priori, which is elaborated in Section 6.3.2.

A feature vector $\mathbf{f}_{i,t}$ is used to represent each bounding box in terms of its position, motion and shape. This work follows the previous work in [She+13; She+15a] to extract the information of position and motion. A challenge in this tracking task results from the similarity of the objects of interest, all of which have a dark appearance, similar shape, and no texture. Therefore, some widely used features (such as colour histogram [HE05], image patch [BF06] and Haar-like features [ZZY12]) are not good choices for discriminative representation here. For example, the advantage of point based features originates from the discriminative local appearance at interest points [Low04; MA10; Ara12a], which is distinct from surroundings (or other targets)

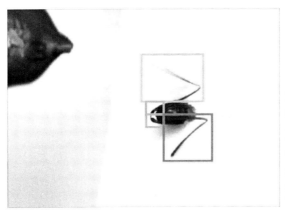

Figure 6.5: The closed up sample video frame of an ant. In the case of an ant, two antennae (yellow and blue bounding boxes) and its mouthparts (purple and green bounding boxes) need to be tracked (i.e. $n = 4$).

and remains consistent over time. However, the local features at interest points of the targets vary dramatically over time, as they tend to move incoherently. It is illustrated in Figure 6.6 how the Kanade-Lucas-Tomasi (KLT) Feature Tracker [LK81] fails to track the left antenna. The initial interest points are detected by a corner detector [HS88].

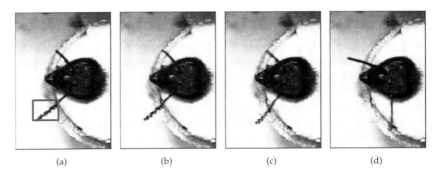

(a)   (b)   (c)   (d)

Figure 6.6: The initial interest points in (a) (Denoted by blue stars) are detected by a corner detector within the green bounding box. The number of successfully tracked points reduces dramatically over time (b) ten, (c) six and (d) zero.

To characterize the shape of each object, an appropriate shape descriptor should be used to model its appearance. The appearance model based on shape context has been successfully used in many machine vision tasks such as frontal face recognition [AZ11], and smooth object retrieval [Ara12b]. The top-hat filter is used as a line detector in the previous work [She+13] to differentiate the honey bee's antenna from other objects, as the bee's antenna is line-shaped. But this is not applicable for other insects such as ants. Popular shape descriptors include EHD [FG11], IEHD [Li+13b], BGF [Li99], SSH [ZL04], FD [ZL03] and ISH [Li+13b]. These six feature extraction methods describe shapes from different perspectives.

To select an effective shape feature for describing the insect body parts, there are two characteristics that should be considered here: First, all body parts have small sizes in the frames (about 200 ~ 600 pixels); second, all body parts have simple bendability, i.e. little local boundary information such as curvature and junctions is present. Some examples of the detected body parts are illustrated in Figure 6.7. The first characteristic limits the discrete points on the edges of the organs (about 50 ~ 200 points), so that the contour-based shape descriptors are not able to obtain enough good sample points. The FD also suffers from this fact. The second characteristic weakens the descriptive power of BGF. By further analysis, this work finds that the body parts' shapes embody good linear edges in different orientations, and because the edges are important low-level features in image description, EHD is chosen as the shape descriptor. EHD is one of the most frequently used contour-based features, which is able to describe both local and global features (see Section 5.3.1), so it is used to describe the global shape features of insect organs.

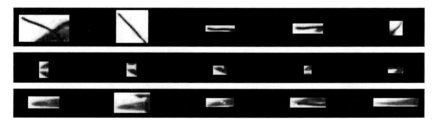

Figure 6.7: The first row are detected antennae, the second row are detected mandibles, and the third row are detected proboscis.

6.3 Interactive Object Tracking Framework

Similar to many DAT approaches (e.g. [ZLN08]), the data association is defined as a MAP problem. The objective is to determine the correspondence of multiple bounding boxes through N frames. Under the MAP framework, a global optimization $\widehat{\mathbf{Y}}_{1:N}$ is found by maximizing the posterior probability $P(\mathbf{Y}_{1:N}|\mathbf{Z}_{1:N})$:

$$\widehat{\mathbf{Y}}_{1:N} = \arg\max_{\mathbf{Y}_{1:N}} P(\mathbf{Y}_{1:N}|\mathbf{Z}_{1:N}) = \arg\max_{\mathbf{Y}_{1:N}} \prod_{t=1}^{N} P(Z_t|Y_t)P(\mathbf{Y}_{1:N}) \quad , \qquad (6.1)$$

where Z_t, Y_t are ordered collections of bounding boxes $\mathbf{z}_{i,t}$ and its label $y_{i,t}$ at time t: $Z_t = \{\mathbf{z}_{i,t}|1 \le i \le n_t\}, Y_t = \{y_{i,t}|1 \le i \le n_t\}$. $P(Z_t|Y_t)$ is the likelihood of the collection of bounding boxes Z_t being generated from the sequence of labels Y_t, it is assumed to be temporally independent. $P(\mathbf{Y}_{1:N})$ is the a prior probability of a labeling sequence $\mathbf{Y}_{1:N}$. The labels are initially estimated at frame level (Section 6.3.1), and then temporal correlation is considered for refinement by data association (Section 6.3.2).

6.3.1 Object Classification

Due to the symmetry of an insect's appearance, a detection response $\mathbf{z}_{i,t}$ is first classified as one of m classes $c_{i,t}$, where $m \in \{1: \text{antenna}; 2: \text{mandible}; 3: \text{proboscis}\}$. Its label $y_{i,t}$ is estimated by differentiating the details (either on the left hand side or the right hand side) in the following tracking step.

In this work, the SVM is selected as a classifier. It improves the performance of the previous work in [She+14] due to its advantage of dealing with high-dimensional data (see Section 3.5). Above all, an MSVM classifier using 1vs1 strategy is adopted.

Multi-class SVM

SVMs are usually used as efficient classifiers for high-dimensional data, but the original SVMs can only be used for two-class classification. To this end, several MSVM approaches are developed to solve the multi-class classification problem, e.g. 1vs1, 1vsR and DAG methods [PCST00].

Because the 1vs1 (one-versus-one) version of MSVM has a low cost of computational complexity, it is selected in this work. In this method, the class of an image is determined based on a voting strategy of two-class SVMs, each of which is built using a pair of all considered classes $\{\omega_1, \omega_2, \ldots, \omega_L\}$. Thus, if there are L classes in total, $L(L-1)/2$ two-class SVM classifiers have to be used. A visual illustration is shown in Figure 6.8 to explain this 1vs1 voting process, where an image is first classified using all these ten two-class SVMs, then, the final classification result is determined by counting to which class the image has been assigned most frequently.

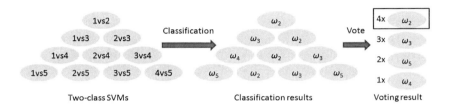

Figure 6.8: An example of the voting process of 1vs1 MSVM, where ω_2 obtains the highest voting number of 4 and is identified as the final classification result.

Classification Using MSVM

The object classification using MSVM mentioned above generates a class label $c_{i,t}$ and the corresponding class probability $P(c_{i,t}|\mathbf{z}_{i,t})$ for each bounding box. [2] Given the output of this classification step, however, two challenges remain in the following tracking task: 1) incorrect classification hypotheses, 2) identity swapping due to the interaction of moving objects.

6.3.2   Constrained Frame-to-Frame Linking

Based on the output of object classification $c_{i,t}$, the appearance information of an insect is exploited, i.e. position and ordering of $\mathbf{z}_{i,t}$, to assign the label $y_{i,t}$. Because the likelihood $P(Z_t|Y_t)$ is assumed to be temporally independent, the label $y_{i,t}$ is determined by the class label $c_{i,t}$ and the relative position of $\mathbf{z}_{i,t}$ to the origin (left or right).

Incorporating Prior Knowledge of the Appearance Model

The likelihood $P(Z_t|Y_t)$ is estimated following the constraint that Z_t should be ordered in an ascending manner, as the body parts of insects are assumed to be ordered in a certain sequence. The label sequences Y_t that violate this assumption will be considered as incorrect hypotheses (i.e. $P(Z_t|Y_t) = 0$). For other Y_t, the likelihood $P(Z_t|Y_t)$ is computed considering the rule of combination without repetition, as n_j bounding boxes are detected out of n objects.

$$P(Z_t|Y_t) = \begin{cases} 0 & \text{if } \hat{m}_1 > m_1 \quad \text{or} \quad \hat{m}_2 > m_2 \\ & \text{or } \hat{m}_3 > m_3 \quad \text{or} \quad \exists \mathbf{z}_{i,t} > \mathbf{z}_{k,t} \quad and \quad \forall k < i \\ \binom{n}{n_j} & \text{otherwise} \end{cases} \quad (6.2)$$

---

[2]Here, the label $c_{i,t}$ is a special format of the above $\{\omega_1, \omega_2, \ldots, \omega_L\}$ in this object tracking system

where m_k is the number of $\{C_t|c_{i,t} = k\}$. This is considered as prior knowledge incorporating the characteristics of insects' appearance. It is easily adapted to other insects by setting the value of m_k and n.

Estimation of Benchmark Frames

The frames with the highest posterior probability are assumed to be correct hypotheses. Among these frames, a set of frames Ψ is defined as the *Benchmark Frames*: Y_b, where $b \in \Psi : P(Z_t|Y_t) = 1 \ \& \ P(Z_{t\pm1}|Y_{t\pm1}) \neq 1$.

$P(\mathbf{Y}_{1:N})$ is defined in Eq. 6.1 to guarantee that only the benchmark frames are used to help rectify the potentially incorrect hypotheses on their neighboring frames by data association:

$$P(\mathbf{Y}_{1:N}) = \prod_{b\in\Psi} P(Y_{b\pm1}|Y_b) \quad . \tag{6.3}$$

The conditional probability $P(Y_{b\pm1}|Y_b)$ is defined as a function of the pair-wise linking cost between Y_b and $Y_{b\pm1}$:

$$P(Y_{b\pm1}|Y_b) = \prod_{i,k} P(y_{i,b} \mapsto y_{k,b\pm1}) \quad , \tag{6.4}$$

where the sign '$\mapsto$' denotes correspondence. The frame-to-frame linking between Y_b and $Y_{b\pm1}$ is found by forming a $n_t \times n_t$ cost matrix $\mathbf{M} = \{M_{i,k}\}$ with

$$M_{i,k} = -logP(y_{i,b} \mapsto y_{k,b\pm1}) = \left\| \mathbf{z}_{i,b} - \mathbf{z}_{k,b\pm1} \right\| \quad , \tag{6.5}$$

where $n_t = \max(n_b, n_{b\pm1})$ and the sign '$\mapsto$' denotes correspondence. As an association optimization algorithm, the Hungarian algorithm [Mun57] is applied to find the optimal linking by minimizing the linking cost.

The likelihood of frames $Y_{b\pm1}$ (i.e. those frames that are rectified with Y_b) is recomputed as

$$P(Z_{b\pm1}|Y_{b\pm1}) = \begin{cases} 0 & \text{if} \quad \hat{m}_1 > m_1 \quad \text{or} \quad \hat{m}_2 > m_2 \\ & \text{or} \quad \hat{m}_3 > m_3 \quad \text{or} \quad \exists \mathbf{z}_{i,b\pm1} > \mathbf{z}_{k,b\pm1} \quad , \quad \forall k < i \quad . \\ 1 & \text{otherwise} \end{cases} \tag{6.6}$$

New benchmark frames are estimated and frame-to-frame linking is performed iteratively.

6.3.3 Interactive KF Estimation and Annotation Query

According to Eq. 6.1 and (6.3), $\widehat{\mathbf{Y}}_{1:N}$ is the current optimal estimation for the labels given a set of benchmark frames in $\{Y_b, b \in \Psi\}$ estimated in Section 6.3.2. The success of frame-to-frame linking lies in the estimation of benchmark frames. Prior knowledge in Eq. 6.2 is used to initially estimate the set of benchmark frames Ψ, but the limitation results from the constraints in Eq. 6.2 do not always hold, and some frames could not be rectified with the given benchmark frames.

To refine further $\widehat{Y}_{1:N}$, it is required to determine new benchmark frames $\{Y_b, b \in \Psi\}$ in Eq. 6.3 to form a new set Ψ^* by introducing human effort. With the new benchmark frames, the constraint in Eq. 6.2 is relaxed. To minimize user effort, an approach is proposed to minimize the number of KFs while optimising the final hypothesis. The intuitive concept is that only the potential benchmark frames should be rectified, so that corrections on the rectified KFs could propagate to their neighboring frames. Given the new set Ψ^* with added KFs obtained from the user annotation, Eq. 6.1 and 6.3 are combined and defined as a new cost function

$$\widehat{Y}^*_{1:N} = \arg\max_{Y^*_{1:N}} \prod_{t=1}^{N} P(Z_t|Y_t) \prod_{b\in\Psi^*} P(Y_{b\pm1}|Y_b) \quad , \tag{6.7}$$

where the refined labels $\widehat{Y}^*_{1:N}$ are found by solving Eq. 6.7.

As illustrated in Figure 6.3, the incorrect hypotheses in $\widehat{Y}_{1:N}$ is refined by an interactive learning framework: 1) requesting user correction on estimated KFs; 2) taking corrected KFs and rectifying their neighboring frames by frame-to-frame linking and 3) updating KFs. The annotation cost of each frame is defined to indicate the degree of "usefulness" of user annotation, in order to estimate which frame should be the potential benchmark frame and added to form a new set of benchmark frames Ψ^*. The higher the annotation cost is, the more erroneous Y_t tends to be. Naturally, the annotation cost is related to the probability of an incorrect hypothesis. Here, two conditions of frames $\widehat{Y}_{1:N}$ are considered, i.e. $Y_{b\pm1}$ and the others. For $Y_{b\pm1}$, we should also take their association with Y_b into consideration. Therefore, the annotation cost is defined as

$$A(Y_t) = P_\epsilon \begin{cases} 1 - P(Z_t|Y_t) \prod_{i,k} P(y_{i,t} \mapsto y_{k,t\pm1}) & t = b \pm 1 \\ 1 - P(Z_t|Y_t) & \text{otherwise} \end{cases} . \tag{6.8}$$

As $A(Y_t)$ interprets the probability that $y_{i,t}$ could be incorrect hypotheses, it provides a flexible strategy for users to set the threshold τ, for which one could choose KFs from the frames $A(Y_t) \geq \tau$ considering the trade-off between tracking accuracy and human effort. The KFs Y_s are defined as $s \in \Phi : P(X_{s-1}|Y_{s-1}) = 1 \,\&\, A(Y_s) \geq \tau$. Users are queried to rectify the KFs Y_s, which are subsequently used to form a new set of benchmark frames as $\Psi^* = \Psi \cup \Phi$.

6.3.4 Track Linking Through Merge Conditions

Given reliable tracklets $\widehat{\mathbf{Y}^*}_{1:N}$ as benchmarks, they are treated as a rough approximation of the tips of each object. To further extract the positions of the tips of each object at pixel level $\mathbf{x}_t^i$ through merge conditions, an approach is proposed to link the tracklets by interpolating the missing tracklets on the in-between frames. Denote the track of the ith target as a set of tracklet association $T^i_{t^p_{i1}, t^p_{i2}} = \{\mathbf{x}_t^i | t^p_{i1} \leq t \leq t^p_{i2}\}$, where t^p_{i1}, t^p_{i2} indicate the tail and head of the pth tracklet of $T^i_{t^p_{i1}, t^p_{i2}}$, respectively.

$$t = t^p_{31} \qquad\qquad\qquad t = t^p_{32}$$

Figure 6.9: An example of linking tracks through merge condition: the shaded lines indicate the tracks at bounding box level, and the circles indicate the tips. t^p_{31}, t^p_{32} indicate the tail and head of the pth tracklet of the proboscis (i.e. label 3) $T^3_{t^p_{31}, t^p_{32}}$, respectively.

For the merge condition where tips of targets a and b are merged (i.e. they are bounded within the same bounding box labeled $y_{a,t}$), $P_m(\mathbf{x}_t^a, \mathbf{x}_t^b \xrightarrow{m} y_{a,t})$ is defined to indicate the probability. It is defined as the product of the independent appearance component $P_{a,m}(\mathbf{x}_t^a, \mathbf{x}_t^b \xrightarrow{m} y_{a,t})$ and the temporal component $P_{t,m}(\mathbf{x}_t^a, \mathbf{x}_t^b \xrightarrow{m} y_{a,t})$, respectively.

$$P_m(\mathbf{x}_t^a, \mathbf{x}_t^b \xrightarrow{m} y_{a,t}) = P_{a,m}(\mathbf{x}_t^a, \mathbf{x}_t^b \xrightarrow{m} y_{a,t})P_{t,m}(\mathbf{x}_t^a, \mathbf{x}_t^b \xrightarrow{m} y_{a,t}) \quad, \tag{6.9}$$

where

$$P_{a,m}(\mathbf{x}_t^a, \mathbf{x}_t^b \xrightarrow{m} y_{a,t}) = \begin{cases} 1 & \text{if } \mathbf{f}_{a,t} \in \Xi \\ 0 & \text{otherwise} \end{cases}, \tag{6.10}$$

and

$$P_{t,m}(\mathbf{x}_t^a, \mathbf{x}_t^b \xrightarrow{m} y_{a,t}) = \begin{cases} 1 & \text{if } t^p_{b1} < t < t^p_{b2} \\ 0 & \text{otherwise} \end{cases}. \tag{6.11}$$

Here, Ξ is a set of $\mathbf{f}_{a,t}$ that constrains the position and size of the target. The starting and ending time indices t^p_{i1}, t^p_{i2} of the pth track $T^i_{t^p_{i1}, t^p_{i2}}$ are empirically set by defining the gap between its temporal neighboring tracks larger than a threshold α, i.e. $t^p_{i2} < t^{p+1}_{i1} - \alpha$.

The estimated tracks are initialised as the set of confident tracklets $\mathbf{T}^0 = \{\mathbf{x}_t^a | P_m(\mathbf{x}_t^a, \mathbf{x}_t^b \xrightarrow{m} y_{a,t}) = 0\}$. The tip $\mathbf{x}_t^a$ is determined by applying morphological operations: the object is firstly thinned to lines, and the end point that is furthest to the centroid of the insect's head is estimated as the tip.

To fill the frame gap under merge condition, Harris corner detector is used to find M candidate pixel positions $\mathbf{x}_t^i, 1 \leq i \leq M$ to interpolate detection responses for estimating $\mathbf{x}_t^a$ and $\mathbf{x}_t^b$. The set of candidate points is denoted as $\{\mathbf{x}_t^i \in \Omega\}$. The estimated tracks are constructed with newly added points selected from Ω, which have the least pair-wise linking costs to their temporally nearest neighbours in $\mathbf{T}^0$.

In summary, the overall algorithm is shown in Algorithm 1 and 2.

Algorithm 1 Summary of the proposed algorithm (Sub-task 1).

    Assign $y_{i,t}$ for each $\mathbf{z}_{i,t}$

Input: $\{\mathbf{z}_{i,t}\}, n, m_k$

Output: $\widehat{\mathbf{Y}}^*_{1:N}, A(Y_t)$

    1. Initialization: For each frame Z_t, compute $P(Z_t|Y_t)$ following Eq. 6.2.

    2. Updating:

    while $\exists\, Y_t$ *updated* **do**

    end while

    for $t = 1, \ldots, N$ **do**

    end for

- Find the benchmark frames $\{Y_b\}$, where $b \in \Psi : P(Z_t|Y_t) = 1$ & $P(Z_{t\pm1}|Y_{t\pm1}) \neq 1$.

- Apply pair-wise linking only on $\{Y_b, b \in \Psi\}$ and their temporal neighbors $Y_{b\pm1}$, update labels $Y_{b\pm1}$.

- Mark $Y_b, Y_{b\pm1}$ *updated*.

    3. KF estimation and annotation query:

- Query user correction and receive correction $Y_s, s \in \Phi : P(X_{s-1}|Y_{s-1}) = 1$ & $A(Y_s) \geq \tau$.

- Form a new set of benchmark frames as $\Psi^* = \Psi \cup \Phi$.

- Update $P(Z_s|Y_s) = 1, \forall s \in \Phi$.

    4. **if** $\Phi \neq \varnothing$ **then**

    repeat step 2-3

    end if

6.4   Summary

In this chapter, an interactive learning framework aiming to achieve a high precision in tracking multiple targets is proposed by minimizing additional human effort for correction. This method integrates a frame query approach, enabling users to correct the erroneous tracking hypotheses and making full use of the user input to optimize

---

Algorithm 2 Summary of the proposed algorithm (Sub-task 2).

Find the tip position $\mathbf{x}_t^i$ and link tracks through merge conditions

Input: $\{\mathbf{z}_{i,t}\}, n, m_k$

Output: $\mathbf{T}$

1. Initialization:

   - Construct initial tracks $\mathbf{T}^0 = \{\mathbf{x}_t^i \in \Omega\}$.

   - Estimate t_{i1}^p, t_{i1}^p as the tail and head of the pth track $T_{l_{i1}}^p$ by empirically setting a threshold of the gap between neighboring tracks α, i.e. $t_{i2}^p < t_{i1}^{p+1} - \alpha$.

2. Updating:

   for $t = t_{i1}^p, \ldots, t_{i2}^p$ **do**

       if $\exists \mathbf{x}_t^i \in \Omega$ at time t **then**

           for $\epsilon = -1, +1, \ldots, -\alpha, +\alpha$ **do**

               if $\exists \mathbf{x}_{t+\epsilon}^a$ or $\mathbf{x}_{t+\epsilon}^b \in \mathbf{T}^0$ at time $t + \epsilon$ **then**

                   - Set $\mathbf{x}_{t+\epsilon}^a, \mathbf{x}_{t+\epsilon}^b$ as the nearest temporal neighbors.
                   - Apply pair-wise linking only on $\{\mathbf{x}_t^i \in \Omega\}$ and their nearest temporal neighbors $\mathbf{x}_{t+\epsilon_a}^a, \mathbf{x}_{t+\epsilon_b}^b \in \mathbf{T}^0$, determine $\mathbf{x}_t^a, \mathbf{x}_t^b$.
                   - Update current tracks $\mathbf{T}$ by $\mathbf{T} = \mathbf{T}^0 \bigcup \{\mathbf{x}_t^a, \mathbf{x}_t^b\}$.

               end if

           end for

       end if

   end for

---

the final results. This is a preferable approach to the traditional track-and-then-rectification scheme, as it does not require an additional round of manual evaluation and correction while guaranteeing a high precision of the tracking results. Particularly, an important aspect of this system's advantage is that it is able to estimate the trajectories of insect body parts at pixel precision even in merge conditions. The practicability and tracking performance of this system is validated on challenging micro-alike video datasets for insect behavioral experiments.

Chapter 7

Applications and Experiments

In this chapter, the CBMIA methods proposed in the former chapters are examined to prove their usefulness and effectiveness, including image segmentation, feature extraction, strongly and weakly supervised learning, unsupervised learning, object tracking and more. In particular, these methods are mainly tested on three microbiological tasks: Environmental microorganism (EM) classification in Section 7.1, stem cell analysis in Section 7.2 and honey bee tracking in Section 7.3. After this work, some additional experiments are implemented in Section 7.4 for extending these CBMIR methods to wider and more generic fields. Finally, a brief conclusion summarises this chapter in Section 7.5.

7.1   Environmental Microorganism Classification

In this section, EM classification is implemented based on various experimental settings to test the functions of different CBMIA approaches. First, Section 7.1.1 introduces the applied dataset of EM images. Then, the proposed semi-automatic image segmentation (see Section 3.2.2), global and local shape feature extraction (see Section 3.3 and 3.4), and strongly supervised learning (SSL) approaches (see Section 3.5) are tested in Section 7.1.2. Finally, the usefulness of sparse coding (SC) feature (see Section 4.2) and weakly supervised learning (WSL) (see Section 4.3) algorithms are proved in Section 7.1.3.

7.1.1   EM Dataset

Before this work, no EM image datasets were specially prepared for examining the CBIMA approaches. Therefore, an EM dataset (EMDS) containing four versions (EMDS-1, EMDS-2, EMDS-3 and EMDS-4) is designed and established to meet this research blank, which contains both classified original EM images and their ground truth images (the manually segmented images). [1] EMDS-1 and EMDS-2 both contain ten classes of EMs. Each class is represented by 20 microscopic images. The difference of EMDS-1 and EMDS-2 is that EMDS-1 is built based on the species of EMs, but EMDS-2 is based on the species and subspecies of EMs, where EMDS-2 contains more detailed information than EMDS-1. EMDS-3 includes 15 classes of EMs, where each class contains 20 images. EMDS-4 is constituted by 21 classes of EMs and each class has 20 images. These different versions of EMDS are used for different sub-tasks in EM classification. The detailed information of EM categories in EMDS is given in Table 7.1, and an example of EMDS is shown in Figure 7.1.

Table 7.1: Information of EM categories.

	EM Categories
EMDS-1	*Actinophrys, Arcella, Aspidisca, Codosiga, Colpoda, Epistylis, Euglypha, Paramecium, Rotifera, Vorticella*
EMDS-2	*Actinophrys, Arcella, Aspidisca, Codosiga, Colpoda, Epistylis, Euglypha, Paramecium, Rotifera, Vorticella*
EMDS-3	*Actinophrys, Arcella, Aspidisca, Codosiga, Colpoda, Epistylis, Euglypha, Paramecium, Rotifera, Vorticella, Noctiluca, Ceratium, Stentor, Siprostomum, Keratella Quadrala*
EMDS-4	*Actinophrys, Arcella, Aspidisca, Codosiga, Colpoda, Epistylis, Euglypha, Paramecium, Rotifera, Vorticella, Noctiluca, Ceratium, Stentor, Siprostomum, Keratella Quadrala, Euglena, Gymnodinium, Gonyaulax, Phacus, Stylongchia, Synchaeta*

Especially because EMDS-3 is the most mature one among all four EMDS versions, containing the second most data and already having been tested over a long period of time (nearly one year), it is selected and used for the following experiments in Section 7.1.2 and 7.1.3.

[1]EMDS information: EMDS-1 was proposed in March, 2012. The original images were acquired in the Civil and Environmental Engineering School of the University of Science and Technology Beijing (USTB), China. The ground truth images are supported by the Research Group for Pattern Recognition in the University of Siegen (Uni-Siegen), Germany. EMDS-2 was modified based on EMDS-1 in September 2012, which improved the quality of the EM microscopic images. EMDS-3 was an extension of EMDS-2, which was finished in October 2013 and it covered more EM classes. EMDS-4 is the newest version of EMDS, which extends the scale of EMDS-3. The extension of this new version is supported by the Chengdu University of Information Technology (CUIT), China in February 2015.

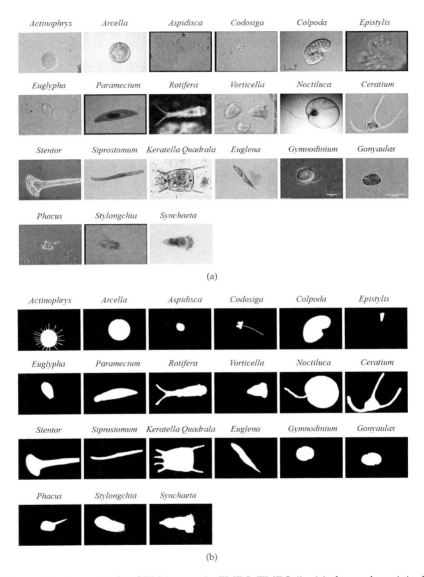

Figure 7.1: An example of EM images in EMDS (EMDS-4). (a) shows the original images. (b) shows the ground truth images.

7.1.2   Classification Using SSL

Evaluation of Segmentation

The necessity of image segmentation for EM classification is first assessed. To this end, the following two types of classification results are compared. The first type is obtained using features extracted from original images. In the second type, features are extracted from ground truth images, each of which is created by manually delineating an EM region. Using such ground truth images, the effect of wrongly detected EM regions is avoided. It should be noted that since the extraction of SIFT feature does not require image segmentation, it is excluded in this experiment.

Figure 7.2 shows the comparison between classification results on original images and those on ground truth images. From the second to the the last row, the representations of APs for 15 classes of EMs are shown, where different marks are used depending on features and used images. The top row presents means of APs for 15 classes ($\omega_1, \omega_2, \ldots, \omega_{15}$). As can be seen in Figure 7.2, while all APs based on original images are around 10% to 30%, APs based on ground truth images range from 30% to 80%. This validates the necessity of image segmentation. In particular, BGF based on ground truth images achieves the best mean AP of 87.31%, which is 67.11% higher than BGF based on original images. ISH using ground truth images leads to the mean AP 57.45%, which shows a great improvement compared to 14.56% in the case of original images. The third one, IEHD, yields the mean AP of 33.59%, and the relative improvement is 14.11%. Finally, the mean AP of FD is increased from 8.27% using original images to 31.71% using ground truth images. The aforementioned results indicate that image segmentation is necessary to achieve accurate EM classification.

Now, the proposed semi-automatic segmentation method in Section 3.2.2 is evaluated. Classification results based on the proposed method are compared to the ones achieved for images segmented by two baseline segmentation methods. The first is a full-automatic method where Sobel edge detector and morphological operations are used to detect candidate regions in an image before the candidate region with the largest area is selected as the EM region. The second baseline method is GrabCut which is a popular semi-automatic segmentation method [BVZ01]. In GrabCut, a user firstly chooses an interesting region in an image as the foreground region, and the remaining region is regarded as the background. Then, these foreground and remaining regions are iteratively refined by modelling the colour distribution in each region with a Gaussian Mixture Model (GMM), and assigning each pixel to the foreground or background. Thereby, the foreground and background regions are characterised by their own distinctive colour distributions.

Figure 7.3 shows the comparison among classification results using the full-automatic, GrabCut, and the proposed semi-automatic segmentation methods. These

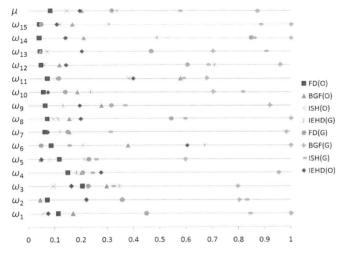

Figure 7.2: EM classification results on original and ground truth images, where 'O' and 'G' on the right side of feature names represent that features are extracted from original and ground truth images, respectively.

results are obtained using BGF and ISH features, which are extracted from EM regions obtained by the above-mentioned segmentation methods. Similar to Figure 7.2, APs are plotted with different marks depending on segmentation methods and features. As shown in Figure 7.3, BGF and ISH based on the proposed semi-automatic method yields more accurate classification results than those based on the full-automatic method and GrabCut. Specifically, the mean AP of BGF using the proposed semi-automatic method is 38.8% and 12.66% higher than those using the full-automatic method and GrabCut, respectively. Also, for ISH, the proposed method leads to the mean AP which is 16.27% and 3.69% higher than those by the full-automatic method and GrabCut, respectively. For IEHD and FD, results have been also obtained which show the effectiveness of the proposed semi-segmentation method. However, performances using IEHD and FD are significantly lower than those using BGF and ISH. Thus, for a clear visualisation of the classification results, the results by IEHD and FD in Figure 7.3 are omitted. The results above verify the effectiveness of the proposed semi-automatic segmentation method for EM classification.

Next, a detailed comparison among the full-automatic, GrabCut and the proposed semi-automatic segmentation methods is conducted. To do so, three evaluation measures are applied, including similarity Si, sensitivity Se, and specificity Sp. These are computed by comparing the segmentation result of an image to the corresponding ground truth. Si describes the similarity between them, Se measures the proportion of

Figure 7.3: EM
classification
results   on
BGF and ISH,
where 'F', 'GC'
and 'S' on the
right   side   of
feature   names
represent   that
features   are
extracted from
segmentation
results by the
full-automatic,
GrabCut   and
the   proposed
semi-automatic
methods.

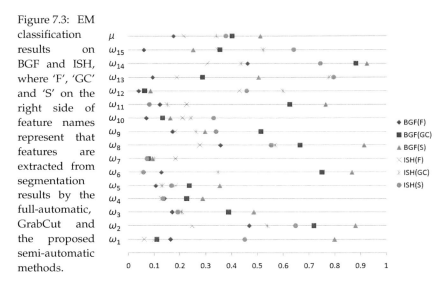

relevant pixels that are correctly labelled by a segmentation method, and *Sp* expresses
the proportion of irrelevant pixels correctly labelled by the method.

Table 7.2 shows the evaluation results of the full-automatic, GrabCut and the
proposed semi-automatic methods in terms of *Si*, *Se* and *Sp*. As can be seen in Table 7.2,
the proposed method has higher performances than the full-automatic method on *Si*
and *Se*, but has a lower performance on *Sp*. These results show the proposed method
has a better overall performance than the full-automatic method. The reason of lower
Sp of the proposed method is that the full-automatic method obtains many total
failed segmentation results, where the whole segmented image is considered as the
background, leading to a high *Sp* value. In Table 7.2, people also find that the proposed
method outperforms GrabCut for all the three evaluation measures. Furthermore,
people can see the proposed method and GrabCut have similar performances (*Si*
= 97.7% → 96.0%, *Se* = 98.8% → 98.4%, and *Sp* = 98.8% → 97.4%), because both of
them involve manual region selection steps. But, since the proposed method has
an additional step to confirm the final EM region, its performance exceeds that of
GrabCut. Figure 7.4 shows segmentation results of three EM images by the full-
automatic, GrabCut and the proposed semi-automatic methods. Due to the low
contrast and strong noise in original images, the full-automatic method delivers poor
segmentation results. As shown in the bottom row in Figure 7.4, a lot of noise is
left by GrabCut, while it is removed by the confirmation step in the proposed semi-

automatic method. Thus, this confirmation step which only requires a single cursor click improves the segmentation performance and also yields the improvement of the EM classification performance shown in Figure 7.3.

Table 7.2: Evaluation of segmentation results in terms of similarity Si, sensitivity Se, and specificity Sp (given in [%]). 'F', 'G' and 'S' in parenthesis represent the full-automatic, GrabCut and the proposed semi-automatic methods, respectively. μ indicates the mean value over 15 classes.

	ω_1	ω_2	ω_3	ω_4	ω_5	ω_6	ω_7	ω_8	ω_9	ω_{10}	ω_{11}	ω_{12}	ω_{13}	ω_{14}	ω_{15}	μ
Si(F)	95.3	84.9	98.8	98.4	92.0	98.4	97.2	91.9	92.5	95.6	71.3	90.0	87.9	94.8	85.7	91.7
Si(G)	90.7	98.7	99.4	97.7	98.4	99.2	97.9	98.1	97.5	99.1	96.7	96.0	96.1	93.6	80.5	96.0
Si(S)	97.2	98.8	99.7	98.9	96.3	99.7	98.9	97.2	97.6	99.1	95.5	96.1	96.4	97.4	96.8	97.7
Se(F)	96.1	84.9	98.8	98.6	92.2	99.1	97.7	92.5	92.6	95.8	71.4	90.0	88.1	95.0	85.8	91.9
Se(G)	98.1	99.6	99.8	99.5	98.9	99.4	99.2	98.5	98.6	99.5	98.7	97.8	98.3	99.7	90.7	98.4
Se(S)	98.1	99.6	99.9	99.7	98.7	99.8	99.7	98.4	99.2	99.6	96.2	97.9	98.1	98.9	98.9	98.8
Sp(F)	99.8	100	99.9	99.8	99.8	99.2	99.4	99.1	100	99.8	99.8	99.9	99.7	99.8	99.9	99.7
Sp(G)	92.3	99.1	99.6	98.2	99.4	99.8	98.7	99.5	98.9	99.5	97.8	98.2	97.8	93.8	88.8	97.4
Sp(S)	99.1	99.2	99.8	99.2	97.5	99.9	99.3	98.6	98.4	99.6	99.2	98.2	98.3	98.5	97.9	98.8

Evaluation of Features

In this chapter, the performances of five selected features (IEHD, BGF, FD, ISH and SIFT) are assessed. Especially perimeter (BGF_1), area (BGF_2), complexity (BGF_3), major axis length (BGF_7), minor axis length (BGF_8) and elongation of approximate ellipse (BGF_9) are chosen based on the preliminary test. Figure 7.5 shows APs and mean APs of these features on ground truth images, where they are plotted with different marks depending on the features. Here, BGF is the most discriminative and ISH comes second. Although BGF obtains the best mean AP of 87.31%, it loses effectiveness for some classes. For example, BGF fails to capture the 'acorn' shape of ω_{10} and the 'awl' shape of ω_{13} (see Figure 7.1). On the other hand, ISH can describe ω_{10}'s shape with bulges and ω_{13}'s shape which broadens gradually, based on angles defined by points on these shapes' counters. However, as shown in Figure 7.6 which shows performances of five features on results by the proposed semi-automatic segmentation method, global features (IEHD, BGF, FD, and ISH) significantly depend on segmentation qualities. For example, BGF's performance on segmentation results is 36.08% lower than that on ground truth images. In addition, ISH's performance

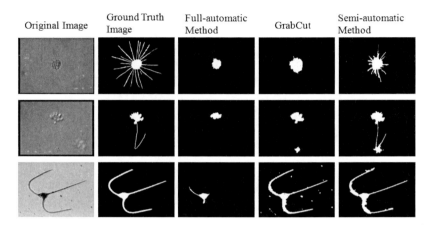

Figure 7.4: Examples of segmentation results. From left to right, each row shows the original and ground truth images of EMs (*Aspidisca*, *Codosiga* and *Ceratium*), the segmentation results by the full-automatic method, the results by GrabCut, and the results by the proposed semi-automatic method.

becomes 19.58% lower. Nonetheless, ISH is the second most discriminative feature among global ones. Apart from global features, as shown in Figure 7.5 and 7.6 SIFT feature achieves a moderate EM classification performance without using image segmentation. The next section will show that global features extracted from imperfect segmentation results greatly contribute to accurate EM classification.

Evaluation of Fusion

In this part, the late fusion scheme described in Section 3.5.3 is evaluated for EM classification. To this end, the proposed late fusion approach is compared to two baseline methods: the first is early fusion that concatenates all five feature vectors into a high-dimensional vector and builds an SVM on it. Furthermore, to avoid the bias caused by different value ranges of different features, each feature is normalised, so that its maximum and minimum values become 1 and 0, respectively. The second baseline method is multiple kernel learning (MKL) which combines the kernel for each feature into a single decision function [Xu+10]. Specifically, MKL with the following

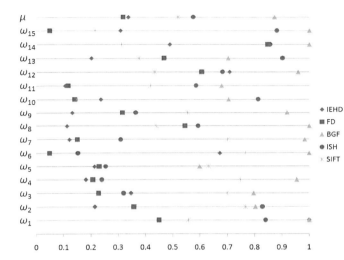

Figure 7.5: EM classification results of five features on ground truth images.

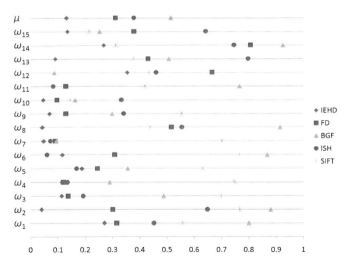

Figure 7.6: EM classification results of five features on results by the proposed semi-automatic segmentation method.

decision function is used:

$$\theta(C) = \sum_{i=1}^{|C|} \beta_i \sum_{j=1}^{H} w_j^i K_i(\mathbf{c}_j^i, \mathbf{c}^i) \quad , \tag{7.1}$$

where C represents an image described by the set of five features $\{\mathbf{c}_{\text{IEHD}}, \mathbf{c}_{\text{BGF}}, \mathbf{c}_{\text{FD}}, \mathbf{c}_{\text{ISH}}, \mathbf{c}_{\text{SIFT}}\}$, so $|C| = 5$ kernels are used. In addition, $\mathbf{c}^i$ represents the ith feature of the image. $K_i(\mathbf{c}_j^i, \mathbf{c}^i)$ represents the kernel value between the image and the jth training image on the ith feature. w_j^i is the weight associated with the $\mathbf{c}_j^i$ in a single kernel, and β_i is the fusion weight for the ith kernel. Thus, the decision function in Eq. 7.1 linearly combines all $|C|$ kernels by their corresponding weights β_i. In this work, the method [Xu+10] which obtains the optimal w_j^i and β_i by iterating the following two processes is used: The first one finds kernel weights β_i which best combines current single kernel functions characterised by fixed w_j^i, and the other process updates parameters of single kernel functions w_j^i based on fixed β_i.

Using Figure 7.7, the late fusion method is first compared to the early fusion and MKL methods. In order to evaluate the effectiveness of fusion methods independently on segmentation errors, these methods use features extracted from ground truth images. The late fusion method achieves the mean AP of 95.97%. Compared to this, the mean AP of the early fusion is 51.74%. For the MKL method with the mean AP of 7.42%, it suffers from the over-fitting problem caused by the condition where positive images are significantly outnumbered by negative ones (see Section 2.1.4 for a detailed discussion about over-fitting). In addition, when numbers of positive and negative images are balanced by randomly selecting the same number of negative images to positive ones, the mean AP of the MKL method becomes 61.33%. However, due to insufficient negative images, the MKL method is still significantly outperformed by the late fusion method. The above-mentioned results validate the suitability of the late fusion method for EM classification.

Furthermore, to assess the effectiveness in practical cases, late fusion on segmentation results by the proposed semi-automatic method is conducted, and its performance is compared to that on ground truth images. From Figure 7.7 people can find that the mean AP on segmentation results is 83.78%, which significantly improves the single features' performances (see Figure 7.6). This performance can be further improved using weights calculated from ground truth images, where the mean AP becomes 88.22%. Because ground truth images do not have segmentation errors, more effective weights can be obtained from them, compared to semi-automatic segmentation results. The result above verifies the practicability of the proposed method which can accurately find a particular EM from many images where various EMs appear.

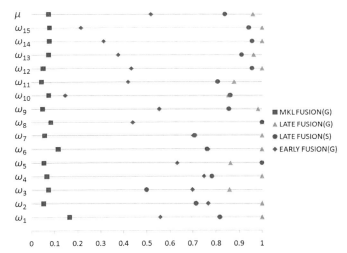

Figure 7.7: EM classification results using early fusion, late fusion and MKL. 'G' means using ground truth images, and 'S' means using the segmentation results by the proposed semi-automatic method.

By comparing Figure 7.6 to Figure 7.7, people can see how late fusion significantly improves classification results. In particular, it yields an impressive performance improvement for ω_5, ω_9 and ω_{10}, which are difficult to recognise only with single features. APs of the late fusion method for these classes amount to 100%, 85.53%, and 86.13%, respectively. Table 7.3 presents fusion weights obtained by applying the late fusion method to semi-automatic segmentation results. This shows that ISH contributes to all classes, which validates its usefulness for EM classification. In addition, because IEHD and FD are edge-based features, they are useful to describe EMs with thin and linear tails, such as ω_{12} (see Figure 7.1). This way, the late fusion method can effectively bring out the synergy of different features to improve the EM classification result.

7.1.3 Classification Using WSL

In the following experiments, ten images from each class of EMDS-3 are randomly selected as training images, and the remaining ones are used for testing. An RBSVM for each class is built by considering training images of this class as positive, and images of all the other classes as negative. A classification result represents the ranking of test images based on scores in Eq. 4.9 and 4.13. Finally, average precision (AP) is used as the evaluation measure of a classification result, and the mean of APs for all EM classes as the overall measure.

Table 7.3: Fusion weights obtained by the proposed late fusion method on semi-segmented EM images.

	ω_1	ω_2	ω_3	ω_4	ω_5	ω_6	ω_7	ω_8	ω_9	ω_{10}	ω_{11}	ω_{12}	ω_{13}	ω_{14}	ω_{15}
IEHD	0.00	0.60	0.00	0.00	0.00	0.00	0.00	0.00	0.00	0.00	0.00	0.30	0.00	0.00	0.00
BGF	0.85	0.20	0.30	0.55	0.50	0.70	0.70	0.00	0.05	0.30	0.00	0.15	0.10	0.30	0.10
FD	0.10	0.15	0.00	0.40	0.45	0.25	0.10	0.00	0.15	0.00	0.15	0.15	0.05	0.45	0.00
ISH	0.05	0.05	0.15	0.05	0.05	0.05	0.20	1.00	0.30	0.70	0.85	0.15	0.15	0.20	0.90
SIFT	0.00	0.00	0.55	0.00	0.00	0.00	0.00	0.00	0.50	0.00	0.00	0.25	0.70	0.05	0.00

Evaluation of NNSC Features

The effectiveness of NNSC features for EM classification is first demonstrated. To this end, the following four types of classification results are compared: ($SVM+NNSC$), ($SVM+BoVW$), ($RBSVM+NNSC$) and ($RBSVM+BoVW$). To make a fair comparison, the same layouts of patches (or interesting points) for both NNSC and BoVW features are used, where patches are densely located with the size of 14-by-14 pixels in an image. The centres of each two neighbouring patches have an interval of 7 pixels. To apply BoVW features, SIFT features are used. One million patches are first randomly selected from the dataset. Then, the k-means clustering method is used to group SIFT features for these patches into 1000 clusters, where each cluster centre is a visual word. For ($SVM+NNSC$), each patch in an EM image is represented using its NNSC weight vector, and the whole image is represented as a 1000-dimensional feature vector, where each dimension represents the sum of weights of patches in the image. For ($SVM+BoVW$), by assigning each patch in an EM image to its most similar visual word, this image is represented as a 1000-dimensional feature vector, where each dimension represents the frequency of a visual word. ($RBSVM+NNSC$) represents each image patch by its NNSC weight vector, and a subwindow is represented by a 1000-dimensional feature vector, where each dimension corresponds to a basis, and represents the sum of weights of patches in the subwindow. ($RBSVM+BoVW$) assigns each patch in a subwindow to its most similar visual word, so that the subwindow is represented as a 1000-dimensional feature vector. Table 7.4 shows a comparison of the classification results of these four methods. The top row displays four types of classification approaches. From the second to the bottom row, the classification results of 15 EM classes are represented by their APs and mean APs. As can be seen in Table 7.4, the mean AP of ($SVM+NNSC$) is higher than that of ($SVM+BoVW$), and the

MAP of ($RBSVM+NNSC$) is also higher than that of ($RBSVM+BoVW$). This validates that NNSC features work more robustly than BoVW features for EM classification.

Table 7.4: EM Classification results using NNSC and BoVW features by linear SVM and RBSVM. APs and the mean of APs (μ) are in [%]. '($S+N$)' shows ($SVM+NNSC$), '($S+B$)' means ($SVM+BoVW$), '($R+N$)' is ($RBSVM+NNSC$) and '($R+B$)' represents ($RBSVM+BoVW$).

	ω_1	ω_2	ω_3	ω_4	ω_5	ω_6	ω_7	ω_8	ω_9	ω_{10}	ω_{11}	ω_{12}	ω_{13}	ω_{14}	ω_{15}	μ
($S+N$)	32.6	55.3	57.4	37.8	42.2	56.9	39.8	35.8	4.4	9.7	38.5	31.0	23.1	30.0	21.4	34.4
($S+B$)	4.0	36.0	65.7	63.4	39.0	44.3	30.5	26.2	29.9	7.4	18.1	19.3	17.1	17.5	12.7	28.7
($R+N$)	37.9	60.7	17.2	40.3	17.9	87.6	11.5	20.5	11.5	17.6	27.5	23.9	17.0	12.7	17.0	28.1
($R+B$)	28.3	53.6	18.3	39.6	16.5	50.3	40.1	35.5	12.6	16.4	8.8	18.7	10.6	15.4	11.5	25.1

Evaluation of Improved RBSVM

First, the usage of the improved RBSVM method from the perspective of localising EMs is proved. Figure 7.8 shows the localisation examples of the improved RBSVM. People can find that this RBSVM can localise most EMs well, but loses effectiveness in some difficult cases. For successful localisation results in Figure 7.8(a), 7.10(c) and 7.8(c), some surrounding areas of the interesting EMs are also localised, because RBSVM identifies them as contextual clues of the EMs. For the difficult localisation cases in Figure 7.8(d), RBSVM identifies only a particular part of an EM as the interesting object, where the classification result may still be correct, but the localisation is incomplete. Because this particular part contains very specific characteristics, only using this part is regarded as leading to the most accurate classification. Figure 7.8(e) shows some transparent EMs, where RBSVM is hard to find their invisible body parts. For these EMs, the transparent body parts have very similar features as the background, so they are localised wrongly.

Next, the usefulness of the improved RBSVM for EM classification is shown. To this end, the classification results between linear SVM and RBSVM are compared. As can be seen from Table 7.4, for both BoVW and NNSC features, the performance using RBSVMs is similar to that using SVMs. Particularly, RBSVM obtains improvements on ω_1, ω_2, ω_4, ω_6, ω_9 and ω_{10}, because these six classes of EMs have semi-transparent bodies and bigger sizes, which contribute more discriminative NNSC features. For ω_{11}, ω_{12}, ω_{13} and ω_{15}, RBSVM has APs similar to those obtained through SVM.

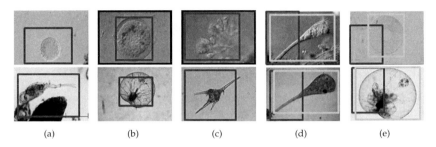

(a)   (b)   (c)   (d)   (e)

Figure 7.8: The localisation results of the improved RBSVM.

However, since the searched region might contain some additional useless areas or lose some useful parts, the remaining classes have lower classification results.

Evaluation of Fusion

However, the four classification results in Table 7.4 are complementary, which offers a possibility to increase the final EM classification result. Therefore, the late fusion is used to enhance the EM classification result like the process in Section 7.1.2. Figure 7.9 shows the classification results of late fusion and the best classification results among the four results in Table. 7.4, where late fusion increases the classification results of all 15 EM classes. The second to bottom row represents APs for 15 classes, where different marks are used depending on different approaches. The top row presents the mean APs for 15 classes. From Figure 7.9, people can find that the late fusion approach increases the classification results of all EM classes and obtains a mean AP of 54.11%, which proves that the improved RBSVM method has a huge potential for cooperating with other methods.

7.2 Stem Cell Analysis and Clustering

In the following experiments, a data set including 81 microscopic images of stem cells acquired in the University of Konstanz, Germany is applied, which contains around 6000 cells (this number is manually calculated by the prior knowledge of the experts). Especially eight ground truth images are prepared by biological experts, which include 104 cells in total.

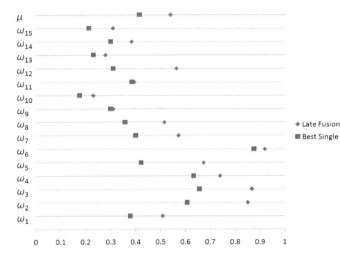

Figure 7.9: EM classification results using late fusion in the WSL framework.

7.2.1 Evaluation of Segmentation

To evaluate the usefulness of the proposed double-stage segmentation, its segmentation result is compared to three baseline methods, including watershed, region growing and thresholding segmentation. The watershed segmentation first uses the nucleus of stem cells as seeds to approximately verify the position of each cell. Then, based on these positions, the cells are segmented by the "watershed" between each two of them. Region growing segmentation first treats each pixel in an image as a cell of a certain type, where these cells can be foreground, background or noise. Then, as the algorithm proceeds, these cells compete to dominate the image domain. Thresholding segmentation first analyses the histograms of the type of images that people want to segment for choosing a proper threshold value. Then, by selecting an adequate threshold value, the grey-level image can be converted into a binary image. Finally, according to the threshold value, all these grey-level values can be classified into two groups (foreground and background). Besides the above mentioned three baseline methods, double-stage segmentation is also compared to two state-of-the-art segmentation methods, namely normalized cut segmentation [SM00] and contour detection and hierarchical image segmentation [Arb+10]. In the normalised cut method, distinctive features are first extracted from all pixels in an image. Then, a graph-based clustering strategy is used to group these pixels into different clusters using the extracted features. Finally, similar clusters are combined as the segmentation result.

Contour detection and hierarchical image segmentation first uses edge detection to find the main parts of each cell. Then, the edge detection is used again to find detailed parts of cells. Lastly, the results of the former two steps are combined by a bottom-up hierarchy. In Table 7.5, the comparison of these methods is shown, which is similar to the format of Table 7.2.

Table 7.5: A comparison of the double-stage segmentation to other methods in terms of similarity Si, sensitivity Se, specificity Sp and ratio Ra of segmented cells and existing cells (given in [%]).

	Si	Se	Sp	Ra
Double-stage	90.75	97.55	66.67	49.83
Watershed	92.33	33.30	81.00	140.62
Region Growing	92.42	21.24	95.24	118.28
Thresholding	84.73	62.69	89.09	85.67
Normalised Cut	92.34	52.11	95.24	118.28
Contour Detection and Hierarchical	92.40	40.00	89.29	128.83

From Table 7.2, people can find that the double-stage segmentation achieves a high Si of 90.75%, and the best Se of 97.55%. However, its Sp is 66.67%, which is rather low compared to other's results. So, the double-stage segmentation has a good overall performance on these three measures. Next, another value is used to compare the segmentation results of these methods, which is the ratio Ra of segmented cells and existing cells in the bottom row of Table 7.2. When Ra is higher, over-segmentation occurs more frequently. Otherwise, under-segmentation occurs more frequently. Hence, the double-stage segmentation has the minimum over-segmentation (2990 cells are segmented from 6000 cells), but suffers from an under-segmentation problem. Furthermore, a visible comparison of the aforementioned segmentation methods is shown in Figure 7.10.

7.2.2   Evaluation of Clustering

Based on the results of the double-stage segmentation, the global features of 2990 segmented stem cells are first extracted. Then, the k-means clustering approach is used

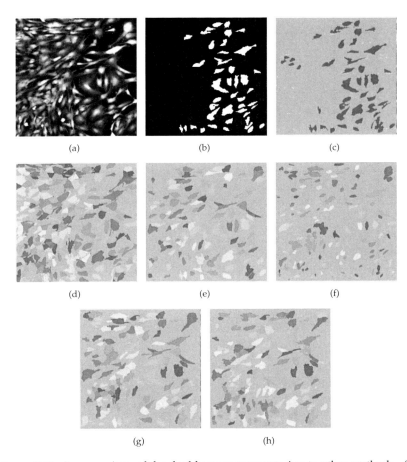

Figure 7.10: A comparison of the double-stage segmentation to other methods. (a) An original image; (b) The ground truth image; (c) Double-stage; (d) Watershed; (e) Region Growing; (f) Thresholding; (g) Normalised Cut; (h) Contour detection and hierarchical image segmentation. Different colours are used to distinguish neighbouring segmented cells.

to group these cells into different clusters using these extracted features [2]. Table 7.6 shows the clustering results that are evaluated by mean variance (see Section 5.4.2). From Table 7.6, people can find that Zernike obtains the highest mean variance 0.8228 when $k = 4$, so four clusters is chosen as the most promising number. Furthermore, when $k = 4$ Hu achieves the second highest value 0.808, which also shows a good clustering performance. Hence, Zernike and Hu are selected for an early fusion approach to enhance the overall clustering result. In the bottom row, the early fusion result of Zernike and Hu is shown, which finally obtains a higher value 0.833 than all the results obtained by single features.

Table 7.6: Mean variance evaluation of stem cell clustering results. The first column shows different global shape features. The first row shows different k numbers.

	4	5	6	7	8	9	10
EHD	0.742	0.714	0.690	0.691	0.673	0.673	0.643
IEHD	0	0	0	0	0	0	0
BGF	0.747	0.698	0.701	0.680	0.667	0.644	0.659
Hu	0.808	0.825	0.826	0.811	0.759	0.707	0.728
Zernike	0.828	0.816	0.736	0.733	0.721	0.715	0.711
SS	0.249	0.229	0.162	0.179	0.164	0.183	0.129
SSH	0.757	0.753	0.753	0.713	0.709	0.673	0.648
FD	0	0	0	0	0	0	0
SCF	0.704	0.645	0.579	0.478	0.477	0.405	0.469
ISH	0.429	0.629	0.592	0.533	0.511	0.502	0.455
Zernike + Hu	0.833	0.811	0.820	0.708	0.721	0.715	0.714

Next, to examine the degree of convergence of the data points (features of stem cells), the silhouette plot is applied (see Section 5.4.2). In Figure 7.11, a silhouette plot is shown, where the early fusion of Zernike and Hu when $k = 4$ is plotted. From Figure 7.11 can be found that these four clusters have a balanced and intensive distribution, which means the clustering process is stable and robust.

---

[2]All 23 components of BGF are used.

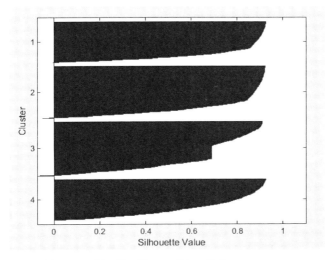

Figure 7.11: Silhouette plot of stem cell clustering.

7.3 Insect Body Parts Tracking

In this section, the tracking system which is proposed in Chapter 6 is examined, using micro-alike shots of honey bees as test objects.

7.3.1 Dataset and Experimental Setting

Each individual insect was imaged using a CCD camera ("FMVU-03MTM/C" point grey), in order to record the head with appended body parts (e.g. proboscis, mandibles and antennae). Stimulus delivery (odor) is monitored by lighting an LED within the field of view of the camera, so that data analysis can be done relative to stimulus delivery. Individual bees are harnessed on a platform, with their heads in fixed positions, but able to move antennae and mouthparts freely. The camera is focused statically on the top of an individual bee. Although it would be possible to record with a high speed camera, this work aims to develop a framework that uses affordable cameras such as web-cams or consumer level cameras, which keeps the data volume low.

This work develops the system *LocoTracker* to implement the proposed algorithm in C++, using OpenCV library version 2.4.8 [3] and tested on an Intel Core2 CPU, 3.00

---

[3]http://www.opencv.org

GHz, with 8 GB RAM. A Qt-based [4] graphical user interface (GUI) is constructed to display KF and take user annotations in order to implement user interaction in Section 6.3.3. To determine the KFs, the threshold of annotation cost is set as $\tau = 1$. The GUI displays each KF and the initial hypotheses, so that the user is able to recognise the errors and correct the mismatches, false negatives and false positives. Figure 7.12 shows two snapshots of the GUI, illustrating how this system facilitates user interactions.

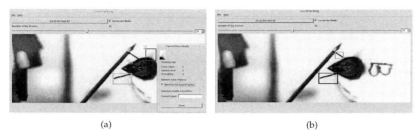

(a)   (b)

Figure 7.12: Two snapshots of the GUI: Initial tracking hypothesis on a KF (a) and user corrected labels (b). *LocoTracker* enables users to correct tracking errors including mismatches, false positives and false negatives.

LocoTracker is tested on recorded videos of two types of insects, i.e. ten videos of a bee and one additional video of an ant. The anatomical model is trained on ten manually annotated objects for each type. The characteristics of the video tested are listed in Table 7.7, including length (number of tested frames), Imaging resolution (pixels per μm), frame rate (frames per second), **GT** (number of ground truth tracks) and **UO** (unobserved objects). Particularly, the ratio of **UO** is measured as $\frac{no.\ of\ frames\ that\ contain\ unobserved\ objects}{no.\ of\ frames}$ to indicate the tracking gaps due to complicated motion patterns of body parts (e.g. the antenna above the insect's head, or the proboscis not extended). The higher the value is, the more tracking gaps the video presents.

Table 7.7: The characteristics of tested videos.

Insect	Length	Imaging Res.(pix/μm)	Frame rate (f/s)	GT	UO
Bee	80000	39	60	5	0.50±0.11
Ant	430	22	30	4	0.13

---

[4]http://qt-project.org/

7.3.2   Experimental Results

LocoTracker is tested in terms of practicability and accuracy. Practicability is measured in two ways: 1) processing time of automated computation and user correction, and 2) the trade-off between human effort and tracking accuracy. Accuracy is validated by comparing the results of the proposed algorithm with some state-of-the-art tracking methods as well as ground truth. Ground truth is manually annotated by a student assistant in the University of Konstanz, Germany.

Practicability

The complexity of the algorithm is measured by processing time. The average running time is recorded for automated computation parts (Section 6.2.1, 6.2.2, 6.3.1 and 6.3.2) and the user correction time. For the running time, it takes about 0.1 seconds per frame. For recording the user correction time, the other student tested LocoTracker finding that it takes about eight seconds to correct all object labels on each KF. Taking the mean of user correction time by averaging over the whole video, it is about 0.8 seconds per frame, thus the additional human labor is tolerable. At each iteration given the user correction at requested KFs, computing Eq. 6.7 takes less than 0.1 seconds. Therefore, the response time of the software between each iteration of user correction is trivial. For comparison, the established software Zootracker [BF14], which also provides user correction, is tested on bee videos. It is designed based on the prior that the displacement between adjacent frames is small and the appearance changes gradually [BF06]. It is a single target tracker which takes about six seconds per object per frame, as user correction is required for most of the video frames.

The trade-off between human intervention and tracking accuracy is tested on bee videos. Figure 7.13(a) shows the convergence of the iterative KF estimation and annotation query (Section 6.3.3). The KF ratio ($KF\ ratio = \frac{no.\ of\ KFs}{no.\ of\ frames}$) depends on the difficulty of tracking: more KFs are estimated for more challenging videos. For all tested videos, the main workload is concentrated in the first five iterations. Figure 7.13(b) shows the accuracy improvement versus the average annotation time at the 0th (before user correction), 1st, 3rd and final iteration. Accuracy is measured by the ratio of tracking errors **TE** (i.e. the number of incorrectly labeled frames) defined in [BJG11]. The **TE** for all bee videos drops below 0.05 at the final interation, while the additional annotation time is about one second on each frame. In summary, the **TE** is 0.02 ± 0.01 for all tested videos, with the user correction only at the KF ratio as 0.14 ± 0.02 and additional annotation time.

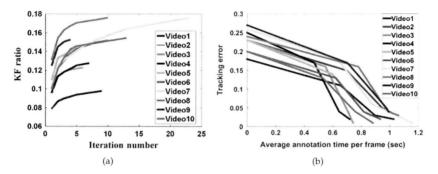

(a)   (b)

Figure 7.13: (a) KF ratio vs. Iteration number of ten tested videos: The user query stops at a KF ratio of 0.1 ~ 0.18, and the KF ratio drops dramatically within five iterations. (b) Tracking error vs. annotation time: The **TE** for all bee videos drops below 0.05 at the final interaction, while additional annotation time is about one second on each frame.

Accuracy

Sub-task 1: The proposed tracking method is compared with several state-of-the-art Association based tracking and Category free tracking methods.

First, the proposed tracking method is compared with the established software Ctrax [Bra+09] and the base tracker [She+15a] that estimates assignment automatically. Ctrax is designed for tracking multiple Drosophila adults, but cannot tackle the situations when the number of targets is not constant and if occlusions are too complex [Del+14]. Identity switch errors occur in the cases of false detection, presence of proboscis and occlusions or merges. Three different methods are tested on one of the bee videos and an ant video for comparison. Ctrax is not applicable for tracking an ant's antennae, as they do not fit the shape prior of Ctrax. The output of a bee video by Ctrax contains only the tracks of two antennae and assumes errors in tracking other body parts, thus only these two targets are taken into account in Table 7.8.

Table 7.8: **TE** of three methods on different insect videos.

	Ctrax [Bra+09]	Base tracker [She+15a]	Proposed system
Bee	0.73	0.10	0.02
Ant	\	0.14	0.02

Second, the state-of-the-art Category free tracking methods (CT [ZZY12], MTT [Zha+12], SPOT [ZM13] and TLD [KMM10]) and the proposed method are tested on the same video. The tested codes are provided by the authors. With the initially annotated right (orange coloured) and left antenna (blue coloured), the tracking results at frames {3, 11, 43} for the first three methods and the proposed method are shown in Figure 7.14. The different tracking methods are denoted using different line types. All of the compared methods start to drift from the right antenna at frame 3, and lose both antennae at frame 43, even when no interaction of targets is present. TLD fails to track the right antenna at the second frame, because valid feature points drop from 100 to 8. Besides, the median of Forward-Backward error is too large (70 pixels). Its detector outputs two bounding boxes with similar confidence, so it terminates both tracking and detection in the following frames. This indicates that Category Free Tracking methods are not applicable for tracking insect body parts from a low frame rate video, as temporal correlation is too weak to predict the position of a target at the current frame given the previous frame.

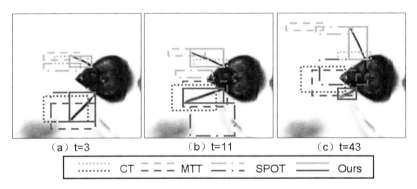

(a) t=3   (b) t=11   (c) t=43

.... CT = = = MTT = . = SPOT ——— Ours

Figure 7.14: Results of four tracking methods.

Sub-task 2: The final tracking results are the position of the tip of each object. Table 7.9 shows results for various flavors of the proposed algorithm. To further show the robustness of the proposed tracking algorithm, the ratio of detection errors is listed after preprocessing described in Section 6.2.1, including *merged detections*, *occluded detections*, *false negatives* (FN) and *false positives* (FP). It can be observed that the estimated position of tips by the proposed approach is very close to ground truth. The mean of position error of all objects is merely five to eight pixels, which is small compared to the average size of the bee's head. The position error of the ant's antennae is relatively higher than the bee's because the motion blur is more severe due to the lower frame rate. Thus, the exact position of the tip is ambiguous.

To show how well the tracks are linked, the method in [Per+06] is carried out, where the track completeness factor **TCF** is used as measurement. A **TCF** of 1 is the ideal, indicating that the final tracks completely overlap with the ground truth. The **TCF** for most objects are above 0.92, except for the proboscis, as it has the highest occluded detection ratio. If the object is occluded, it does not make sense to track its position. This measurement indicates the advantage of the proposed approach in linking tracks in merged conditions, which produces tracks comparable to manual "point and click" results (see Table 7.9).

Table 7.9: Comparison of proposed method with ground truth.

Object Name	Right Antenna	Proboscis	Left Antenna
Average position error (pixels)	5.70±0.58	7.50±0.78	6.00±1.10
TCF	0.92±0.03	0.55±0.18	0.94±0.01
Merged (%)	0.29±0.24	12.60±5.96	0.71±1.14
Occluded (%)	0.13±0.23	12.60±1.81	0.56±0.51
FN (%)	7.80±2.60	3.80±3.00	5.80±1.28
FP (%)	0.35±0.25	0.00±0.00	1.04±0.82

To show the advantage of the proposed algorithm in fulfilling two sub-tasks, ten consecutive sample frames of the final tracking results are illustrated in Figure 7.15. This is an extreme case of merge condition. As the result of sub-task 1, the label of each bounding box is estimated. The correct labels are given in (e) with the help of user correction, even though they do not follow the ascending order. Given the reliably labeled bounding boxes, the positions of proboscis tips in (a)-(i) in merged bounding boxes are estimated with acceptable precision by the proposed track linking approach.

As the final output, three trajectories of tips are drawn on one of the video frames for visualization, as shown in Figure 7.16.

Anatomical Model

To validate the analysis of the anatomical model of insect body parts in Section 6.2.2, the classification results of Section 6.3.1 for six global shape features are listed in

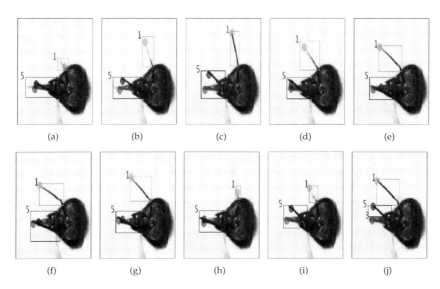

Figure 7.15: Ten consecutive sample frames of the final tracking results under merge condition.

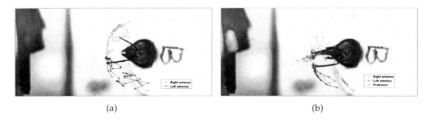

Figure 7.16: Three trajectories of tips of 100 frames in the videos shown in Figure 6.4 are drawn on one of the video frames: The orange dots denote the tips of the right antenna, red for the proboscis and blue for the left antenna.

Table 7.10 tested on bee videos. Similar to most biomedical data, the class distribution is skewed. For example, the number of mandibles is much smaller than the number of antennae. The unbalanced data problem causes the minority class to be more prone to misclassification than the majority class. Taking the unbalanced classes into account, the mean and variance are calculated by treating each class with equal weight. As shown in Table 7.10, EHD obtains the highest mean value of classification rate and the lowest variance, and is therefore selected for this work.

Table 7.10: Classification results in Section 6.3.1 for six global shape features.

	Antenna	Mandible	Proboscis	Mean	Variance
EHD [FG11]	0.96	0.66	0.55	0.72	0.05
IEHD [Li+13b]	0.16	0.43	0.95	0.51	0.16
BGF [Li99]	0.99	0.34	0.67	0.67	0.11
SSH [ZL04]	0.99	0.44	0.43	0.62	0.10
FD [ZL03]	0.44	0.32	1.00	0.59	0.13
ISH [Li+13b]	0.99	0.33	0.47	0.60	0.12

7.4   Extended Experiments

In this chapter, some extended experiments are carried out. Some of them are used to enhance of the persuasion of the proposed CBMIA approaches. The others are the extended applications of the existing methods.

7.4.1   CBMIR on EMDS Images

Image Retrieval

In this work, an microscopic image retrieval approach is proposed which performs a similarity search in an image dataset for a query given by a user. To calculate the similarity, feature vectors of the query and dataset images are extracted first. Then, the Euclidean distance between the query image and each dataset image is calculated. Thirdly, the dataset images are ranked by their similarities in a descending order. Finally, this ranked image dataset is used as the image retrieval result, where the images in former positions are more similar to the query image, and the images in latter positions are more unlike to the query image.

In order to evaluate the performance of an image retrieval, there are some measures often used, e.g. precision and recall. In this work, average precision (AP) and the mean value of AP are used for evaluating the retrieval result, where the definition of AP can be found in Eq. 3.37.

Overview of the System

To increase the search effectiveness of environmental microbiological information, a novel *Environmental Microbiological Content-based Microscopic Image Retrieval* (EM-CBMIR) system is proposed using CBMIA techniques. Given a query image from a user, this system starts by conducting a semi-automatic segmentation process to obtain only the region of interest, which defines the EM without any surrounding artifacts. Then, features which characterise the shape of the EM are extracted from the segmented image. To finalise the system process, the euclidean distance between the extracted features from the query image and features extracted formerly from all images in the database is calculated. This retrieval system outputs similar images to the query sorted from most similar to dissimilar, and gives feedback to the user. The framework of this system is shown in Figure 7.17.

In this work, the following problems are addressed:

- Image Segmentation:
  To ensure the accuracy of removing impurities in microscopic images, the semi-automatic segmentation method which is proposed in Section 3.2.2 is applied.

- Feature Extraction:
  In this system, ISH is used to discriminate structural properties of a microorganism (see Section 3.3.4), in order to show the effectiveness of this newly developed feature extraction method.

- CBMIR Approach:
  The CBMIR method mentioned in Section 7.4.1 is used to search images.

EMDS-4 is used to test the effectiveness of this EM-CBMIR system, which contains 21 classes of EMs $\omega_1, \ldots, \omega_{21}$ (see Section 7.1.1). In the experiments of EM retrieval, each EM image is used as a query image once and all remaining images as database images. AP is used to evaluate the result of such a retrieval process, and the mean of APs for all 21 classes is used as an overall evaluation measure. Finally, to avoid the bias caused by different value ranges of different features, each feature is normalised, so that its maximum and minimum values become 1 and 0, respectively.

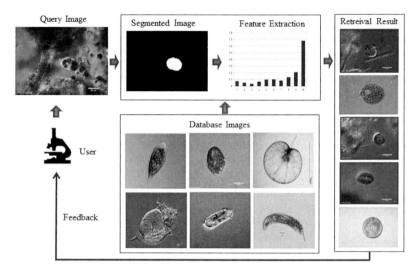

Figure 7.17: Working flow of the proposed EM-CBMIR system.

Evaluation of the System

The usefulness of ISH in EM-CBMIR system is evaluated in Figure 7.18, where the retrieval results of ISH with ground truth images and segmentation results are both shown.

In Figure 7.18, people can find that ISH obtains a mean AP of 33.9% on segmentation results, which is very close to that of ground truth images of 36.8%. This comparison shows the robustness of ISH feature for EM-CBMIR tasks on semi-segmentation results. Furthermore, an example of the retrieval result is shown in Figure 7.19.

7.4.2   Late Fusion of Global Shape Features on MPEG-7 Dataset

In order to examine the generality of the selected global shape features including ISH proposed in Section 3.3, an additional experiment using MPEG-7 shape collection [LLE00] is conducted. This dataset contains 400 images where each of the 20 classes (e.g. cars, keys and watches) includes 20 images. An example of this dataset is shown in Figure 7.20.

In this dataset, the shapes in all images are perfectly segmented. To compare the results of the proposed features to the ones of existing methods, this experiment is conducted in the framework of the 'query by example' retrieval (see Section 7.4.1).

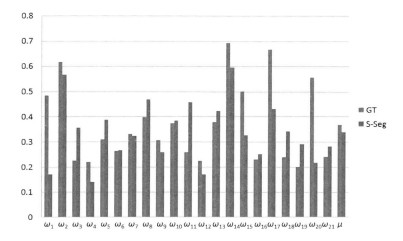

Figure 7.18: Comparison of EM-CBMIR results using ISH on ground truth (GT) and semi-automatic segmented (S-Seg) images. The horizontal axis shows the AP and mean AP (μ). The vertical axis shows the value of AP and mean AP.

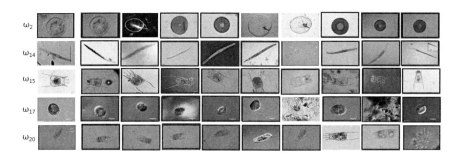

Figure 7.19: Examples of EM-CBMIR results using ISH. The first column shows the query images. From the second to the last column, the database images are sorted by their ISH similarities from high to low. The images in red bounding boxes are the relevant images.

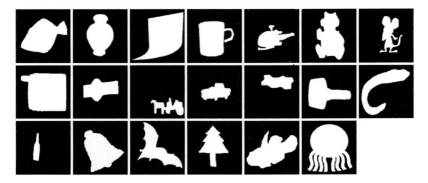

Figure 7.20: An example of the MPEG-7 dataset.

Specifically, one of 400 images is selected as a query image, and then a retrieval result is obtained where the remaining 399 images are ranked based on their similarities to the query image. In the case of this work, for each feature, the similarity between two images is computed as their Euclidean distance. Finally, the result is evaluated by checking whether or not images ranked in the top ten positions are relevant (belong to the same category) to the query image.

Table 7.11 shows performances of the proposed global features on the MPEG-7 dataset. The first column shows feature names. Columns two to eleven show the statistics of images ranked in the first to the 10th position in retrieval results. Given a retrieval result, the counter of each position is incremented if the image ranked at this position is relevant to the query image. Values in Table 7.11 are counter values that are obtained by checking retrieval results using all the 400 images as queries. For example, the row of ISH shows that, in 349 of 400 retrieval results, the top ranked images are relevant to the corresponding query images. The rightmost column presents values of the overall evaluation measure. This represents the average of precisions over 400 retrieval results, where each precision is computed using images ranked in the top ten positions.

Table 7.11 shows that BGF, FD, ISH and IEHD are the best, the second best, the third best and the worst features on MPEG-7 dataset. Except FD, the other three features have similar performance trends to those on EMDS. For FD, in MPEG-7 dataset, most objects are located at the centre of the images. As a result, geometrical centres are placed inside of object regions. This is a suitable condition for FD to describe the shape of an object, using distances between points on its contour and the geometrical centre. Unlike MPEG-7 dataset, geometrical centres for some EMs are located outside of their regions (e.g. ω_{14} in Figure 7.1), which decreases the descriptive power of FD.

Table 7.11: Retrieval results on MPEG-7 dataset using the proposed global shape features and PSSGM.

	1st	2nd	3rd	4th	5th	6th	7th	8th	9th	10th	P@10
ISH	349	316	295	290	264	261	244	209	208	194	65.75%
EHD	223	196	179	146	132	128	117	99	113	94	36.50%
BGF	352	341	341	308	300	291	280	273	264	253	75.08%
FD	357	324	308	284	290	280	262	263	230	223	70.53%
Fusion	388	369	364	348	339	339	328	316	297	285	84.33%
PSSGM	380	371	361	351	344	339	332	320	330	309	85.93%

Thus, the performance of FD significantly depends on datasets. Compared to this, ISH can be considered stable because it is useful for both EM and MPEG-7 datasets.

Finally, performances of the proposed features are compared to the one of a state-of-the-art method, PSSGM [BL08]. For the object shown in each image, PSSGM extracts the skeleton which is a one-dimensional line representation, capturing the geometry and topology of the object. The skeleton is formed by points which are located at centres of the object contour. A part (endpoint) of a skeleton is charactised by its topological relation to the other parts. Based on the similarity in this relation, each part of a skeleton is matched to the one of another skeleton. PSSGM computes the similarity between two images by finding the best part matching in the corresponding skeletons. Since using a single feature is clearly insufficient for achieving the state-of-the-art performance, similarities calculated on different features are fused. Here, for each retrieval, only a query image is available, so it is impossible to perform the fusion weight computation described in Section 3.5.3. Thus, the fused similarity between two images is computed by simply adding their similarities on different features. As can be seen in Table 7.11, the simple fusion achieves a performance which is very close to that of PSSGM. This validates the generality of the proposed features which are effective for both EM and MPEG-7 datasets.

7.4.3   EM Classification Using Pair-wise Local Features

Pair-wise SIFT Feature

Pair-wise SIFT (PSIFT) feature is developed to improve the spatial describing ability of the SFIT feature [MS10]. Because the normal SIFT feature only considers the local characteristics of an image, it loses the information between each two different image regions. To overcome this limitation, PSIFT feature is introduced as follows

$$\text{PSIFT}_{i,j} = [\text{SIFT}_i, \text{SIFT}_j] \quad , \quad \text{in this work} \quad i,j = 1,2,\ldots,1000 \quad . \quad (7.2)$$

The position of $\text{PSIFT}_{i,j}$ is the middle position of SIFT_i and SIFT_j. This simple process of PSIFT solves the spatial representation shortage of SIFT, leading to a 256-dimensional PSIFT feature vector. To use this vector, a BoVW approach is applied, which is similar to that in Section 3.4, but one million PSIFT features are used in the k-means clustering process. Finally, a 1000-dimensional feature vector $\mathbf{c}_{\text{PSIFT}} = (\text{PSIFT}_1, \text{PSIFT}_2, \ldots, \text{PSIFT}_{1000})^{\text{T}}$ is achieved, where each dimension represents the frequency of a pair-wise visual word.

Modified Pair-wise SIFT feature

The above mentioned PSIFT feature faces an image rotation problem. When the image is rotated, the extracted normal SIFT feature is invariable, but the combined $\text{PSIFT}_{i,j}$ and $\text{PSIFT}_{j,i}$ are two totally different feature vectors. To overcome this rotation, a *Modified PSIFT* (MPSIFT) feature is proposed in this work. The MPSIFT feature is defined as follows

$$\text{MPSIFT}_{i,j} = \text{MPSIFT}_{j,i} = \text{SIFT}_i + \text{SIFT}_j \quad , \quad \text{in this work} \quad i,j = 1,2,\ldots,1000 \quad . \quad (7.3)$$

The position of $\text{MPSIFT}_{i,j}$ or $\text{MPSIFT}_{j,i}$ is the middle position of SIFT_i and SIFT_j. The MPSIFT feature eliminates the difference of $\text{PSIFT}_{i,j}$ and $\text{PSIFT}_{j,i}$, leading to a solution for the image rotation problem of PSIFT features. MPSIFT features have the same dimensionality as a SIFT feature, so it has a 128-dimensional feature vector. Then, a BoVW process is used as the one in Section 3.4. Lastly, a 1000-dimensional feature vector $\mathbf{c}_{\text{MPSIFT}} = (\text{MPSIFT}_1, \text{MPSIFT}_2, \ldots, \text{MPSIFT}_{1000})^{\text{T}}$ is obtained, where each dimension represents the frequency of a MPSIFT visual word.

EM Classification Using PSIFT and MPSIFT

In this work, three local features are introduced in Section 3.4, including SIFT, PSIFT and MPSIFT features. The effectiveness of a SIFT feature in CBMIA tasks has already been proved in Section 7.1.2 and 7.1.3 on EM classification tasks. Similarly, to show the usefulness of PSIFT and MPSIFT features in the CBMIA domain, their functions are tested in the same task. For this experiment, EMDS-3 is applied. To obtain pair-wise features, a threshold of seven pixels is used, where each interest point is paired with its neighbouring points less than seven pixels in distance. Considering

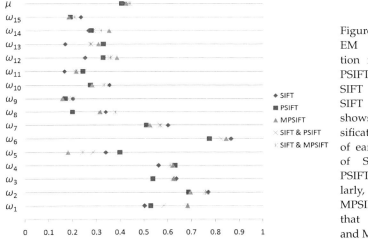

Figure 7.21: EM classification results of PSIFT and MP-SIFT features. SIFT & PSIFT shows the classification result of early fusion of SIFT and PSIFT. Similarly, SIFT & MPSIFT shows that of SIFT and MPSIFT.

the computational cost, a 6-by-6 pixels density of the interest points is used. Finally, the APs of PSIF and MPSIFT are shown in Figure 7.21.

From Figure 7.21, people can gather that PSIFT has a similar performance to SIFT, but MPSIFT obtains a better result than SIFT. Furthermore, a complementarity exists between them, which supports the possibility of feature fusion. Hence, an early fusion approach is applied to enhance the classification result. The classification results of early fusion between SIFT and PSIFT, as well as between SIFT and MPSIFT are shown in Figure 7.21. The results show that early fusion improves the classification results. SIFT & PSIFT obtains 43.4% mean AP which is 1.5% higher than SIFT and 2.8% higher than PSIFT. SIFT & MPSIFT achieves 44.2% mean AP which is 2.4% higher than SIFT and 1.6% higher than MPSIFT. These results underline the pair-wise local features' have a very huge potential in the future work of CBMIA tasks.

7.5   Summary

In this chapter, the proposed CBIMA approaches are examined to prove their effectiveness. In particular, three main microbiological tasks are implemented to test these CBMIA methods, including EM classification in Section 7.1, stem cell analysis in Section 7.2 and insect tracking in Section 7.3. Besides these three main tasks, several additional experiments are carried out in Section 7.4 to extend the CBIMA methods to a wider application domain.

Chapter 8

Conclusion and Future Work

In this last chapter, the present work will be completed by a conclusion. A summary of the proposed CBMIA approaches in this work is given in Section 8.1. Furthermore, Section 8.2 introduces a plan of the future work.

8.1   Conclusion

Microscopic image analysis is very important to microbiology, biological medicine and environmental engineering, which can help researchers to gain more knowledge about a microcosmos. To effectively analyse microscopic images, content-based microscopic image analysis (CBMIA) is developed and applied. Usually, CBMIA addresses the following problems:

- High Noisy Image:
  Noise usually decreases the feature extraction performance of a microscopic image.

- Object Representation:
  To describe information of an object in an image, effective object representation approaches are needed.

- Object Classification and Clustering:
  Object classification and clustering are two important tasks in CBMIA application domains, which can group different objects into relative groups.

- Object Detection and Tracking:
  Object detection and tracking are another two significant goals of CBMIA work, which are able to find and monitor an object and its moving trajectory in a high speed video.

To solve the problems mentioned above, various CBMIA algorithms and techniques are developed and selected in this work as follows:

- High Noisy Image:
  To remove the noise from an image, two main approaches are proposed. The first is Sobel edge detection based image segmentation, which can divide interesting objects from the original image and is useful for increasing the efficiency of feature extraction. Especially a semi- and a full-automatic image segmentation methods are developed. The semi-automatic approach is introduced in Section 3.2, containing two manual operations. The experimental results in Section 7.1.2 shows that this segmentation method has a high performance of Si = 97.7%, Se = 98.8% and Sp = 98.8%. In contrast, the full-automatic method does not need any manual operations, which is more suitable for the segmentation of large scale images (see Section 5.2). In the experiment in Section 7.2.1, this approach finally obtains a good performance of Si = 90.75%, Se = 97.55% and Sp = 66.67%, which underlines its usefulness.

  The second is a weakly supervised learning (WSL) method, which can localise the interesting objects with bounding boxes and exclude (or reduce) the noise around the objects (see Section 4.3). The experimental result in Section 7.1.3 shows that although WSL loses effectiveness to localise objects in some difficult cases, for example small objects, transparent objects and partial objects, it can localise most objects well. Hence, this WSL method has very huge potential in the future CBIMA work.

- Object Representation:
  To effectively describe an object, three types of feature extraction methods are introduced, including global shape, local shape and sparse coding (SC) features. Global shape features represent the object from the overall shape characteristics, like ISH feature (see Section 3.3). The effectiveness of them are proved by the experiments in Section 7.1.2, 7.2.2, 7.3.2, 7.4.1 and 7.4.2.

  Local shape features describe the object from its partial shape properties, like SIFT feature (see Section 3.4). Their usefulness is shown by the experiments in Section 7.1.2 and 7.1.3. Especially, to use the spatial information between each two local features to improve the describing ability, pair-wise local features are proposed and tested in Section 7.4.3, where the effectiveness is shown.

  In contrast to the above-mentioned shape features, SC features can convert raw pixel values into higher-level semantic features, which is effective to represent the object in a more intelligent way (see Section 4.2). The usefulness of SC features is examined in Section 7.1.3, where the performance of SC features observably exceeds that of BoVW features.

- Object Classification and Clustering:
  To classify objects, several classifiers are built in this work, including linear SVM, RBFSVM (see Section 3.5), MSVM (see Section 6.3.1) and RBSVM (see Section 4.3). Specifically, an improved RBSVM approach is inserted in the WSL framework to classify and localise the object jointly (see Section 4.3.2). The effectiveness of linear SVM is improved in Section 7.1.3, where it is inserted into a WSL framework and shows reasonable performance of 25.1% mean AP using NNSC features. The RBFSVM is used in Section 7.1.2 within a SSL framework and obtains good classification result of 95.97% mean AP using a late fusion approach. The usage of MSVM is examined in Section 7.3.2, where a mean classification rate of 0.72 is obtained. The improved RBSVM uses NNSC features instead of the traditional BoVW features and achieves an observable improvement, where it reaches a mean AP of 28.1%. Furthermore, when the late fusion method is applied within the WSL framework, the mean AP improves to 54.11%.

  Besides the supervised learning approaches mentioned above, unsupervised learning is also used in CBMIA work. Because the high generality k-means clustering algorithm, it is applied to effectively group unlabeled images into different groups (see Section 5.4). Its effectiveness is shown by the experiment in Section 7.2.2, where a high mean variance of 0.833 is obtained.

- Object Detection and Tracking:
  To effectively detect and track objects in a video, a multi-object tracking system is established in this work using an interactive learning framework (see Chapter 6), where a frame-to-frame link approach is developed to enhance the tracking performance. In the experiment in Section 7.3, the great usefulness of this method is proved, where multiple objects can be well tracked.

Furthermore, to prove the usefulness and effectiveness of the CBMIA approaches mentioned above, three microbiological applications using these methods are implemented. The first is environmental microorganism (EM) classification, where two systems are built. One is a semi-automatic system using SSL framework (see Section 7.1.2). The other is a full-automatic system using WSL framework, (see Section 7.1.3). The second is stem cell analysis, which uses image segmentation, global shape feature extraction and k-means clustering methods (see Section 7.2). The third is honey bee tracking, where the head organs of a bee in a micro-alike video are detected, tracked and classified using frame-to-frame tracking, global shape features and an MSVM classifier (see Section 7.3).

8.2   Future Work

In the future, in order to further improve the CBMIA approaches, the following respects will be considered and carried out:

- Automatic Image Segmentation:
  To more effectively increase the intelligent degree of the segmentation method, a *Supervised Normalised Cut* (SNC) algorithm is planned to be employed [Yan+13]. This method uses a supervised learning process to enhance the segmentation performance of the original normalised cut algorithm, which would be useful for the complex microscopic images.

  In the plan of SNC segmentation for microscopic images, manually segmented images (ground truth images) are used as training data in a supervised learning framework, where these images first "teach" the machine to know which regions in the images are useful areas. Then, based on the learnt prior knowledge, a basic normalised cut algorithm is applied to group regions in a newly input image into different clusters and obtains the final segmentation result. Because SNC uses a supervised learning approach, it has more sufficient "understanding" of the segmentation task and can do more accurate work than the basic method.

- Pair-wise Sparse Coding Feature:
  The current used SC features only consider the information of each single image patch, but lose the spatial dependency of different patches. To this end, *Pair-wise Sparse Coding* (PSC) feature is planned to be used [MS11a], which can extract more spatial information for describing images. The experiments of PSIFT and MPSIFT features in Section 7.4.3 also prove the significance of the spatial information for describing an image.

  For implementing PSC, two possible approaches are currently in planning. The first is to extract *Pair-wise Bases*, where each two normal bases are grouped into one pair for combining a higher-level basis. The second is to extract *Pair-wise Sparse Weight Vectors*, where each two normal sparse weight vectors are combined into one feature vector.

- Transformation Invariant Sparse Coding Feature:
  Because the normal NNSC feature is sensitive to image rotation, *Transformation Invariant Sparse Coding* (TISC) feature is employed [MS11b] for enhancing the

performance of RBSVM. Since a non-negative condition is sometimes needed, *Non-negative TISC* (NNTISC) in particular is planned as a special version of TISC.

- Extension of EMDS:
  To improve the usefulness of EMDS, EMDS-5 is in planning, where more classes of EMs will be included. Additionally, the corresponding ground truth images will be manually segmented by biological experts. To this end, some appropriate cooperators in following research fields are considered, for example environmental science, environmental engineering, microbiology and ecology.

Bibliography

[AB+00]   J. Alvarez-Borrego et al. 'Invariant Optical Colour Correlation for Recognition of Vibrio Cholerae 01'. In: *International Conference on Pattern Recognition*. 2000, pp. 283–286.

[ADF10]   B. Alexe, T. Deselaers, and V. Ferrari. 'What Is An Object?' In: *IEEE International Conference on Computer Vision and Pattern Recognition*. 2010, pp. 73–80.

[AG12]   H. C. Akakin and M. N. Gurcan. 'Content-based Microscopic Image Retrieval System for Multi-image Queries'. In: *IEEE Transactions on Information Technology in Biomedicine* 16.4 (2012), pp. 758–769.

[AL12]   H. Azizpour and I. Laptev. 'Object Detection Using Strongly-supervised Deformable Part Models'. In: *European Conference on Computer Vision*. 2012, pp. 836–849.

[AZ11]   R. Arandjelovic and A. Zisserman. 'Smooth Object Retrieval Using A Bag of Boundaries'. In: *International Conference on Computer Vision*. 2011, pp. 375–382.

[Ara12a] O. Arandjelovic. 'Contextually Learnt Detection of Unusual Motion-based Behaviour in Crowded Public Spaces'. In: *Computer and Information Sciences II*. 2012, pp. 403–410.

[Ara12b] O. Arandjelovic. 'Gradient Edge Map Features for Frontal Face Recognition under Extreme Illumination Changes'. In: *British Machine Vision Association Conference*. 2012, pp. 1–11.

[Arb+10] P. Arbelaez et al. 'Contour Detection and Hierarchical Image Segmentation'. In: *IEEE transactions on Pattern Analysis and Machine Intelligence* 33.5 (2010), pp. 898–916.

[BF06]   A. Buchanan and A. Fitzgibbon. 'Interactive Feature Tracking Using K-D Trees and Dynamic Programming'. In: *IEEE International Conference on Computer Vision and Pattern Recognition*. 2006, pp. 626–633.

[BF14]   A. Buchanan and A. Fitzgibbon. *Zoo Tracer*. 2014. URL: http://research. microsoft.com/zootracer.

[BJG11]   C. R. del Bianco, F. Jaureguizar, and N. Garcia. 'Bayesian Visual Surveillance: A Model for Detecting and Tracking A Variable Number of Moving Objects'. In: *IEEE International Conference on Image Processing*. 2011, pp. 1437–1440.

[BKV01]   T. Balch, Z. Khan, and M. Veloso. 'Automatically Tracking and Analyzing the Behavior of Live Insect Colonies'. In: *International Conference on Autonomous Agents*. 2001, pp. 521–528.

[BL08]   X. Bai and L. J. Latecki. 'Path Similarity Skeleton Graph Matching'. In: *IEEE Transactions on Pattern Analysis and Machine Intelligence* 30.7 (2008), pp. 1282–1292.

[BMP02]   S. Belongie, J. Malik, and J. Puzicha. 'Shape Matching and Object Recognition Using Shape Contexts'. In: *IEEE Transactions on Pattern Analysis and Machine Intelligence* 24.4 (2002), pp. 509–522.

[BU04]   E. Borenstein and S. Ullman. 'Learning to Segment'. In: *European Conference on Computer Vision*. 2004, pp. 315–328.

[BVZ01]   Y. Boykov, O. Veksler, and R. Zabih. 'Efficient Approximate Energy Minimization via Graph Cuts'. In: *IEEE Transactions on Pattern Analysis and Machine Intelligence* 23.11 (2001), pp. 1222–1239.

[BWG07]   B. Bose, X. Wang, and E. Grimson. 'Multi-Class Object Tracking Algorithm That Handles Fragmentation and Grouping'. In: *IEEE Conference on Computer Vision and Pattern Recognition*. IEEE. 2007, pp. 1–8.

[Bal+92]   H. W. Balfoort et al. 'Automatic Identification of Algae: Neural Network Analysis of Flow Cytometric Data'. In: *Journal of Plankton Research* 14.4 (1992), pp. 575–589.

[Ban13]   P. Bankhead. *Analyzing Fluorescence Microscopy Images with ImageJ*. Queen's University Belfast, UK. 2013.

[Bei+06]   C. Beisbart et al. 'Extended Morphometric Analysis of Neuronal Cells with Minkowski Valuations'. In: *The European Physical Journal B* 52 (2006), pp. 531–546.

[Ber+95]   D. Bernhard et al. 'Phylogenetic Relationships of the Nassulida within the Phylum Ciliophora Inferred from the Complete Small Subunit rRNA Gene Sequences of Furgasonia Blochmanni, Obertrumia Georgiana, and Pseudomicrothorax Dubius'. In: *Journal of Eukaryotic Microbiology* 42.2 (1995), pp. 126–131.

[Bla+98]   N. Blackburn et al. 'Rapid Determination of Bacterial Abundance, Biovol-
            ume, Morphology, and Growth by Neural Network-based Image Analy-
            sis'. In: *Applied and Environmental Microbiology* 64.9 (1998), pp. 3246–3255.

[Blu+11]   M. Blum et al. 'On the Applicability of Unsupervised Feature Learning for
            Object Recognition in RGB-D Data'. In: *NIPS Workshop on Deep Learning
            and Unsupervised Feature Learning.* 2011, pp. 1–9.

[Bra+09]   K. Branson et al. 'High-throughput Ethomics in Large Groups of
            Drosophila'. In: *Nature Methods* 6.6 (2009), pp. 451–457.

[Bur98]   C. J. C. Burges. 'A Tutorial on Support Vector Machines for Pattern Recog-
            nition'. In: *Data Mining and Knowledge Discovery* 2.2 (1998), pp. 121–167.

[CH67]   T. M. Cover and P. E. Hart. 'Nearest Neighbor Pattern Classification'. In:
            IEEE Transactions on Information Theory 13.1 (1967), pp. 21–27.

[CJW02]   H. D. Cheng, X. H. Jiang, and J. Wang. 'Colour Image Segmentation Based
            on Homogram Thresholding and Region Merging'. In: *Pattern Recognition*
            35.2 (2002), pp. 373–393.

[CL07]   L. Cao and F. Li. 'Spatially Coherent Latent Topic Model for Concurrent
            Segmentation and Classification of Objects and Scenes'. In: *IEEE Interna-
            tional Conference on Computer Vision.* 2007, pp. 1–8.

[CL11]   C. Chang and C. Lin. 'LIBSVM: A Library for Support Vector Machines'.
            In: *ACM Transactions on Intelligent Systems and Technology* 3.2 (2011), pp. 1–
            27.

[CRM03]   D. Comaniciu, V. Ramesh, and P. Meer. 'Kernel-based Object Tracking'. In:
            IEEE Transactions on Pattern Analysis and Machine Intelligence 25.5 (2003),
            pp. 564–577.

[CT92]   P. W. Cox and C. R. Thomas. 'Classification and Measurement of Fungal
            Pellets by Automated Image Analysis'. In: *Biotechnology and Bioengineer-
            ing* 39.9 (1992), pp. 945–952.

[Can86]   J. Canny. 'A Computational Approach to Edge Detection'. In: *IEEE Trans-
            actions on Pattern Analysis and Machine Intelligence* 8.6 (1986), pp. 679–698.

[Cao+08]   F. Cao et al. 'Transcriptional and Functional Profiling of Human Em-
            bryonic Stem Cell-derived Cardiomyocytes'. In: *PLOS One* 3.10 (2008),
            e3474.

[Cha+06]   M. A. Chabaud et al. 'Olfactory Conditioning of Proboscis Activity in
            Drosophila Melanogaster'. In: *Journal of Comparative Physiology A* 192.12
            (2006), pp. 1335–1348.

[Chi+11]   L. K. Chin et al. 'Protozoon Classifications Based on Size, Shape and Re-
           fractive Index Using On-Chip Immersion Refractometer'. In: *International
           Solid-state Sensors, Actuators and Microsystems Conference*. 2011, pp. 771–
           774.

[Cla89]   J. J. Clark. 'Authenticating Edges Produced by Zero-crossing Algorithms'.
           In: *IEEE Transactions on Pattern Analysis and Machine Intelligence* 11.1
           (1989), pp. 43–57.

[Cou+09]   J. Cousty et al. 'Watershed Cuts: Minimum Spanning Forests and the
           Drop of Water Principle'. In: *IEEE Transactions on Pattern Analysis and
           Machine Intelligence* 31.8 (2009), pp. 1362–1374.

[Csu+04]   G. Csurka et al. 'Visual Categorization with Bags of Keypoints'. In: *ECCV
           Workshop on Statistical Learning in Computer Vision*. 2004, pp. 1–22.

[Cul+03]   P. F. Culverhouse et al. 'Expert and Machine Discrimination of Marine
           Flora: a Comparison of Recognition Accuracy of Field-collected Phyto-
           plankton'. In: *International Conference on Visual Information Engineering*.
           2003, pp. 177–181.

[Cul+94]   P. F. Culverhouse et al. 'Automatic Categorisation of Five Species of
           Cymatocylis (Protozoa, Tintinnida) by Artificial Neural Network'. In:
           Marine Ecology Progress Series 107 (1994), pp. 273–280.

[Cul+96]   P. F. Culverhouse et al. 'Automatic Classification of Field-collected Di-
           noflagellates by Artificial Neural Network'. In: *Marine Ecology Progress
           Series* 139.1–3 (1996), pp. 281–287.

[DJJ94]   M. Dubuisson, A. K. Jain, and M. K. Jain. 'Segmentation and Classification
           of Bacterial Culture Images'. In: *Journal of Microbiological Methods* 19.4
           (1994), pp. 279–295.

[DLW05]   R. Datta, J. Li, and J. Z. Wang. 'Content-based Image Retrieval: Ap-
           proaches and Trends of the New Age'. In: *ACM SIGMM International
           Workshop on Multimedia Information Retrieval*. 2005, pp. 253–262.

[DLW06]   H. Daims, S. Luecker, and M. Wagner. 'Daime, a Novel Image Analysis
           Program for Microbial Ecology and Biofilm Research'. In: *Environmental
           Microbiology* 8.2 (2006), pp. 200–213.

[DM01]   I. S. Dhillon and D. M. Modha. 'Concept Decompositions for Large Sparse
           Text Data Using Clustering'. In: *Machine Learning* 42.1 (2001), pp. 143–175.

[DMK03]   I. S. Dhillon, S. Mallela, and R. Kumar. 'A Divisive Information-theoretic
           Feature Clustering Algorithm for Text Classification'. In: *Journal of Ma-
           chine Learning Research* 3 (2003), pp. 1265–1287.

[Daz10]   F. B. Dazzo. 'CMEIAS Digital Microscopy and Quantitative Image Analysis of Microorganisms'. In: *Microscopy: Science, Technology, Applications and Education* 2.4 (2010), pp. 1083–1090.

[Del+14]   A. I. Dell et al. 'Automated Image-based Tracking and Its Application in Ecology'. In: *Trends in Ecology & Evolution* 29.7 (2014), pp. 417–428.

[Dre+07]   L. Drever et al. 'Comparison of Three Image Segmentation Techniques for Target Volume Delineation in Positron Emission Tomography'. In: *Journal of Applied Chinical Medical Physics* 8.2 (2007), pp. 93–109.

[EG99]   J. Eakins and M. Graham. *Content-based Image Retrieval: A report to the JISC Technology Applications Programme.* Tech. rep. 39. University of Northumbria at Newcastle, 1999.

[EGH03]   K. V. Embleton, C. E. Gibson, and S. I. Heaney. 'Automated Counting of Phytoplankton by Pattern Recognition: a Comparison with a Manual Counting Method'. In: *Journal of Plankton Research* 25.6 (2003), pp. 669–681.

[Erb+93]   J. Erber et al. 'Antennal Reflexes in the Honeybee: Tools for Studying the Nervous System'. In: *Apidologie* 24.3 (1993), pp. 238–296.

[FCA03]   M. G. Forero, G. Cristbal, and J. Alvarez. 'Automatic Identification Techniques of Tuberculosis Bacteria'. In: *SPIE 5203, Applications of Digital Image Processing.* 2003, pp. 71–81.

[FCD06]   M. G. Forero, G. Cristobal, and M. Desco. 'Automatic Identification of Mycobacterium Tuberculosis by Gaussian Mixture Models'. In: *Journal of Microscopy* 223.2 (2006), pp. 120–132.

[FG11]   H. Frigui and P. Gader. 'Detection and Discrimination of Land Mines in Ground-Penetrating Radar Based on Edge Histogram Descriptors and a Possibilistic k-Nearest Neighbor Classifier'. In: *IEEE Transactions on Fuzzy Systems* 17.1 (2011), pp. 185–199.

[FH89]   T. Fukuda and O. Hasegawa. 'Expert System Driven Image Processing for Recognition and Identification of Micro-organisms'. In: *International Workshop on Industrial Applications of Machine Intelligence and Vision.* 1989, pp. 33–38.

[FM81]   K. S. Fu and J. K. Mui. 'A Survey of Image Segmenation'. In: *Pattern Recognition* 13.1 (1981), pp. 3–16.

[FPZ03]   R. Fergus, P. Perona, and A. Zisserman. 'Object Class Recognition by Unsupervised Scale-invariant Learning'. In: *IEEE Conference on Computer Vision and Pattern Recognition.* 2003, pp. 264–271.

[FV+02]   M. G. Forero-Vargas et al. 'Segmentation, Autofocusing, and Signature Extraction of Tuberculosis Sputum Images'. In: *SPIE 4788, Photonic Devices and Algorithms for Computing IV*. 2002, pp. 1–12.

[Fan+01]   J. Fan et al. 'Automatic Image Segmentation by Integrating Colour-edge Extraction and Seeded Region Growing'. In: *IEEE Transactions On Image Processing* 10.10 (2001), pp. 1454–1466.

[Fel+10]   P. F. Felzenszwalb et al. 'Object Detection with Discriminatively Trained Part Based Models'. In: *IEEE Transactions on Pattern Analysis and Machine Intelligence* 32.9 (2010), pp. 1627–1645.

[Fia+14]   L. Fiaschi et al. 'Tracking Indistinguishable Translucent Objects over Time Using Weakly Supervised Structured Learning'. In: *IEEE Conference on Computer Vision and Pattern Recognition*. 2014, pp. 2736–2743.

[Fri+00]   J. Fried et al. 'Monitoring Protozoa and Metazoa Biofilm Communities for Assessing Wastewater Quality Impact and Reactor Up-scaling Effects'. In: *Water Science and Technology* 41.4–5 (2000), pp. 309–316.

[GB12]   S. H. Gillespie and K. B. Bamford. *Medical Microbiology and Infection at a Glance Fourth Edition*. Hoboken, US: Wiley-Blackwell, 2012.

[GD05]   G. Guo and C. R. Dyer. 'Learning from Examples in the Small Sample Case: Face Expression Recognition'. In: *Systems, Man, and Cybernetics–Part B: Cybernetics* 35.3 (2005), pp. 477–488.

[GE05]   D. Grimaldi and M. S. Engel. *Evolution of the Insects*. Cambridge, UK: Cambridge University Press, 2005.

[GS07]   S. Giannarou and T. Stathaki. 'Shape Signature Matching for Object Identification Invariant to Image Transformations and Occlusion'. In: *Computer Analysis of Images and Patterns*. Ed. by W. G. Kropatsch, M. Kampel, and A.Hanbury. Berlin, Germany: Springer, 2007.

[GSL91]   S. J. Greenwood, M. L. Sogin, and D. H. Lynn. 'Phylogenetic Relationships within the Class Oligohymenophorea, Phylum Ciliophora, Inferred from the Complete Small Subunit rRNA Gene Sequences of Colpidium Campylum, Glaucoma Chattoni, and Opisthonecta Henneguyi'. In: *Journal of Molecular Evolution* 33.2 (1991), pp. 163–174.

[GW08]   R. C. Gonzalez and R. E. Woods. *Digital Image Processing Third Edition*. New Jersey, US: Pearson International Edition, 2008.

[Gal+08]   C. Galleguillos et al. 'Weakly Supervised Object Localization with Stable Segmentations'. In: *European Conference on Computer Vision*. 2008, pp. 193–207.

[Gd10]   F. J. Guerrieri and P. d'Ettorre. 'Associative Learning in Ants: Condition-
           ing of the Maxilla-labium Extension Response in Camponotus Aethiops'.
           In: *Journal of Insect Physiology* 56.1 (2010), pp. 88–92.

[Gee+07]   K. Geels et al. *Metallographic and Materialographic Specimen Preparation,
           Light Microscopy, Image Analysis and Hardness Testing*. US: ASTM Interna-
           tional, West Conshohocken, PA, 2007.

[Gua+12]   N. Guan et al. *MahNMF: Manhattan Non-negative Matrix Factorization*.
           2012. eprint: arXiv:1207.3484.

[HE05]   Y. Huang and I. A. Essa. 'Tracking Multiple Objects Through Occlu-
           sions'. In: *IEEE Conference on Computer Vision and Pattern Recognition*.
           2005, pp. 1051–1058.

[HG09]   H. He and E. A. Garcia. 'Learning from Imbalanced Data'. In: *IEEE Trans-
           actions on Knowledge and Data Engineering* 21.9 (2009), pp. 1263–1284.

[HS88]   C. Harris and M. Stephens. 'A Combined Corner and Edge Detector'. In:
           Alvey Vision Conference. 1988, pp. 147–151.

[HWN08]   C. Huang, B. Wu, and R. Nevatia. 'Robust Object Tracking by Hierarchical
           Association of Detection Responses'. In: *European Conference on Computer
           Vision*. 2008, pp. 788–801.

[HY01]   D. J. Hand and K. Yu. 'Idiot's Bayes–Not So Stupid after All?' In: *Interna-
           tional Statistical Review* 69.3 (2001), pp. 385–398.

[HYT79]   T. Huang, G. Yang, and G. Tang. 'A Fast Two-dimensional Median Fil-
           tering Algorithm'. In: *IEEE Transactions on Acoustics, Speech, and Signal
           Processing* 27.1 (1979), pp. 13–18.

[Hig+15]   T. Higaki et al. 'Semi-automatic Organelle Detection on Transmission
           Electron Microscopic Images'. In: *Scientific Reports* 5 (2015), pp. 1–9.

[Hoy02]   P. O. Hoyer. 'Non-negative Sparse Coding'. In: *IEEE Workshop on Neural
           Networks for Signal Processing*. 2002, pp. 557–565.

[Hu62]   M. Hu. 'Visual Pattern Recognition by Moment Invariants'. In: *IRE Trans-
           actions on Information Theory* 8.2 (1962), pp. 179–187.

[Ish+87]   T. Ishii et al. 'The Identification, Counting, and Measurement of Phy-
           toplankton by an Image-processing System'. In: *ICES Journal of Marine
           Science* 43.3 (1987), pp. 253–260.

[JDM00]   A. K. Jain, R. P. W. Duin, and J. Mao. 'Statistical Pattern Recognition: A
           Review'. In: *IEEE Transactions on Pattern Analysis and Machine Intelligence*
           22.1 (2000), pp. 4–37.

[JLG05]   J. M. Jay, M. J. Loessner, and D. A. Golden. *Modern Food Microbiology Seventh Edition*. New York, US: Springer, 2005.

[JSD10]   Y. Jia, M. Salzmann, and T. Darrell. *Factorized Latent Spaces with Structured Sparsity*. Tech. rep. UCB/EECS-2010-99. Electrical Engineering and Computer Sciences University of California at Berkeley, 2010.

[Jav+05] B. Javidi et al. 'Three-dimensional Imaging and Recognition of Microorganism Using Single-exposure On-line (SEOL) Digital Holography'. In: *Optics Express* 13.12 (2005), pp. 4492–4506.

[Jav+06] B. Javidi et al. 'Real-time Automated 3D Sensing, Detection, and Recognition of Dynamic Biological Micro-organic Events'. In: *Optics Express* 14.9 (2006), pp. 3806–3829.

[Jef+84] H. P. Jeffries et al. 'Automated Sizing, Counting and Identification of Zooplankton by Pattern Recognition'. In: *Marine Biology* 78.3 (1984), pp. 329–334.

[Jia+10] Y. Jiang et al. 'Representations of Keypoint-based Semantic Concept Detection: A Comprehensive Study'. In: *IEEE Transactions on Multimedia* 12.1 (2010), pp. 42–53.

[Jol02]   I. T. Jolliffe. *Principal Component Analysis 2nd Edition*. New York, US: Springer, 2002.

[KB01]   P. Kaewtrakulpong and R. Bowden. 'An Improved Adaptive Background Mixture Model for Real-time Tracking with Shadow Detection'. In: *European Workshop on Advanced Video Based Surveillance Systems*. 2001, pp. 135–144.

[KH90]   A. Khotanzad and Y. H. Hong. 'Invariant Image Recognition by Zernike Moments'. In: *IEEE transactions on Pattern Analysis and Machine Intelligence* 12.5 (1990), pp. 489–497.

[KLW99]   D. M. Kocak, N. D. V. Lobo, and E. A. Widder. 'Computer Vision Techniques for Quantifying, Tracking, and Identifying Bioluminescent Plankton'. In: *IEEE Journal of Oceanic Engineering* 24.1 (1999), pp. 81–95.

[KM08]   S. Kumar and G. S. Mittal. 'Geometric and Optical Characteristics of Five Microorganisms for Rapid Detection Using Image Processing'. In: *Biosystems Engineering* 99.1 (2008), pp. 1–8.

[KM09]   S. Kumar and G. S. Mittal. 'Textural Characteristics of Five Microorganisms for Rapid Detection Using Image Processing'. In: *Journal of Food Process Engineering* 32.1 (2009), pp. 126–143.

[KM10]   S. Kumar and G. S. Mittal. 'Rapid Detection of Microorganisms Using Image Processing Parameters and Neural Network'. In: *Food and Bioprocess Technology* 3.5 (2010), pp. 741–751.

[KMM10]   Z. Kalal, J. Matas, and K. Mikolajczyk. 'P-N Learning: Bootstrapping Binary Classifiers by Structural Constraints'. In: *IEEE Conference on Computer Vision and Pattern Recognition*. 2010, pp. 49–56.

[KR08]   J. Kovacevic and G. K. Rohde. 'Overview of Image Analysis Tools and Tasks for Microscopy'. In: *Microscopic Image Analysis for Life Science Applications*. Ed. by J. Rittscher, R. Machiraju, and S. T. C. Wong. 685 Canton Street, Norwood, MA: Artech House, INC, 2008.

[Kir+00] G. J. Kirkpatrick et al. 'Optical Discrimination of a Phytoplankton Species in Natural Mixed Populations'. In: *Association for the Sciences of Limnology and Oceanography* 45.2 (2000), pp. 467–471.

[Kis05]   K. Kishida. *Property of Average Precision and Its Generalization: An Examination of Evaluation Indicator for Information Retrieval Experiments*. Tech. rep. NII-2005-E014E. National Institute of Informatics, 2005.

[Kla+09] J. Klappstein et al. 'Moving Object Segmentation Using Optical Flow and Depth Information'. In: *Pacific Rim Symposium on Advances in Image and Video Technology*. 2009, pp. 611–623.

[LBH08]   C. H. Lampert, M. B. Blaschko, and T. Hofmaann. 'Beyond Sliding Windows: Object Localization by Efficient Subwindow Search'. In: *IEEE Conference on Computer Vision and Pattern Recognition*. 2008, pp. 1–8.

[LK81]   B. D. Lucas and T. Kanade. 'An Iterative Image Registration Technique with An Application to Stereo Vision'. In: *International Joint Conference on Artificial Intelligence*. Vol. 2. 1981, pp. 674–679.

[LLE00]   L. J. Latecki, R. Lakamper, and U. Eckhardt. 'Shape Descriptors for Nonrigid Shapes with a Single Closed Contour'. In: *IEEE Conference on Computer Vision and Pattern Recognition*. 2000, pp. 424–429.

[LLW07]   H. Lin, C. Lin, and R. C. Weng. 'A Note on Platt's Probabilistic Outputs for Support Vector Machines'. In: *Machine Learning* 68.3 (2007), pp. 267–276.

[LM14]   M. Ladniak and M. Mlynarczuk. 'Search of Visually Similar Microscopic Rock Images'. In: *Computational Geosciences* 19.1 (2014), pp. 127–136.

[LR07]   T. Landgraf and R. Rojas. 'Tracking Honey Bee Dances from Sparse Optical Flow Fields'. In: *FB Mathematik und Informatik FU* (2007), pp. 1–37.

[LRS10]   A. Lemme, R. F. Reinhart, and J. J. Steil. 'Efficient Online Learning of a
           Non-negative Sparse Autoencoder'. In: *European Symposium on Artificial
           Neural Networks*. 2010, pp. 1–6.

[LS10]   S. Lakshmi and V. Sankaranarayanan. 'A Study of Edge Detection Tech-
           niques for Segmentation Computing Approaches'. In: *International Jour-
           nal on Computer Applications, Special Issue on Computer Aided Soft Computing
           Techniques for Imaging and Biomedical Applications* 1 (2010), pp. 35–41.

[LSG15a]   C. Li, K. Shirahama, and M. Grzegorzek. 'Application of Content-based
           Image Analysis to Environmental Microorganism Classification'. In: *Bio-
           cybernetics and Biomedical Engineering* 35.1 (2015), pp. 10–21.

[LSG15b]   C. Li, K. Shirahama, and M. Grzegorzek. 'Environmental Microbiology
           Aided by Content-based Image Analysis'. In: *Pattern Analysis and Appli-
           cations* (2015), Accepted for Publication.

[LSG15c]   C. Li, K. Shirahama, and M. Grzegorzek. 'Environmental Microorganism
           Classification Using Sparse Coding and Weakly Supervised Learning'.
           In: *International Workshop on Environmental Multimedia Retrieval in Con-
           junction with ACM International Conference on Multimedia Retrieval*. 2015,
           pp. 9–14.

[LW09]   D. Lin and X. Wu. 'Phrase Clustering for Discriminative Learning'. In:
           Annual Meeting of the ACL and IJCNLP. 2009, pp. 1030–1038.

[LZK14]   W. Luo, X. Zhao, and T. K. Kim. 'Multiple Object Tracking: A Review'.
           In: *arXiv preprint arXiv:1409.7618* (2014).

[Lee+04]   S. Lee et al. 'Ciliate Populations as Bio-indicators at Deer Island Treatment
           Plant'. In: *Advances in Environmental Research* 8.3–4 (2004), pp. 371–378.

[Lee+07]   H. Lee et al. 'Efficient Sparse Coding Algorithms'. In: *Advances in Neural
           Information Processing Systems*. 2007, pp. 801–808.

[Lee+09]   H. Lee et al. 'Exponential Family Sparse Coding with Application to Self-
           taught Learning'. In: *International Jont Conference on Artifical Intelligence*.
           2009, pp. 1113–1119.

[Li+07]   X. Li et al. 'Local and Global Features Extracting and Fusion for Microbial
           Recognition'. In: *ACIS International Conference on Software Engineering,
           Artificial Intelligence, Networking, and Parallel/Distributed Computing*. 2007,
           pp. 507–511.

[Li+13a]   C. Li et al. 'A Multi-stage Approach for Automatic Classification of Envi-
           ronmental Microorganisms'. In: *International Conference on Image Process-
           ing, Computer Vision, and Pattern Recognition*. 2013, pp. 364–370.

[Li+13b]   C. Li et al. 'Classification of Environmental Microorganisms in Micro-
           scopic Images Using Shape Features and Support Vector Machines'. In:
           IEEE International Conference on Image Processing. 2013, pp. 2435–2439.

[Li99]   S. Z. Li. 'Shape Matching Based on Invariants'. In: *Shape Analysis, Progress
           in Neural Networks*. 1999, pp. 203–228.

[Liu+01]   J. Liu et al. 'CMEIAS: A Computer-aided System for the Image Analysis of
           Bacterial Morphotypes in Microbial Communities'. In: *Microbial Ecology*
           41.3 (2001), pp. 173–194.

[Lon+07]   F. Long et al. 'Phenotype Clustering of Breast Epithelial Cells in Confocal
           Images Based on Nuclear Protein Distribution Analysis'. In: *BMC Cell
           Biology* 8.Suppl 1 (2007), pp. 1–15.

[Low+08]   W. E. Lowry et al. 'Generation of Human Induced Pluripotent Stem Cells
           from Dermal Fibroblasts'. In: *Proceedings of the National Academy of Sciences
           of the United States of America* 105.8 (2008), pp. 2883–2888.

[Low04]   D. G. Lowe. 'Distinctive Image Features from Scale-invariant Keypoints'.
           In: *International Journal of Computer Vision* 60.2 (2004), pp. 91–110.

[MA10]   R. Martin and O. Arandjelovic. 'Multiple-object Tracking in Cluttered and
           Crowded Public Spaces'. In: *Advances in Visual Computing*. 2010, pp. 89–
           98.

[MAA09]   V. Makkapati, R. Agrawal, and R. Acharya. 'Segmentation and Classifica-
           tion of Tuberculosis Bacilli from ZN-stained Sputum Smear Images'. In:
           IEEE International Conference on Automation Science and Engineering. 2009,
           pp. 217–220.

[MC+01]   M. Martin-Cereceda et al. 'Dynamics of Protozoan and Metazoan Com-
           munities in a Full Scale Wastewater Treatment Plant by Rotating Biolog-
           ical Contactors'. In: *Microbiological Research* 156.3 (2001), pp. 225–238.

[MC96]   J. Moreira and L. D. F. Costa. *Neural-based Colour Image Segmentation and
           Classification Using Self-organizing Maps*. Anais do IX SIBGRAPI. 1996.

[MH03]   D. Mara and N. Horan. *Handbook of Water and Wastewater Microbiology*.
           London, UK: Academic Press, 2003.

[MS10]   N. Morioka and S. Satoh. 'Building Compact Local Pairwise Codebook
           with Joint Feature Space Clustering'. In: *European Conference on Computer
           Vision*. 2010, pp. 692–705.

[MS11a]   N. Morioka and S. Satoh. 'Compact Correlation Coding for Visual Object
           Categorization'. In: *IEEE International Conference on Computer Vision*. 2011,
           pp. 1639–1646.

[MS11b]   M. Morup and M. N. Schmidt. 'Transformation Invariant Sparse Coding'. In: *IEEE International Workshop on Machine Learning for Signal Processing*. 2011, pp. 1–6.

[MVD10]   A. Mendez-Vilas and J. Diaz. *Microscopy: Science, Technology, Applications And Education*. Spain: Formatex Research Center, 2010.

[Mac+96]   M. D. Mackey et al. 'CHEMTAX-A Program for Estimating Class Abundances from Chemical Markers: Application to HPLC Measurements of Phytoplankton'. In: *Marine Ecology Progress Series* 144 (1996), pp. 265–283.

[Mac67]   J.B. MacQueen. 'Some Methods for Classification and Analysis of Multivariate Observations'. In: *Berkeley Symposium on Mathematical Statistics and Probability*. 1967, pp. 281–297.

[Mai+10]   J. Mairal et al. 'Online Learning for Matrix Factorization and Sparse Coding'. In: *Journal of Machine Learning Research* 11.42 (2010), pp. 19–60.

[Mar04]   J. M. Martinez. *MPEG-7 Overview (Version 10)*. Tech. rep. ISO/IEC JTC1/SC29/WG11. 2004.

[Mik+05]   K. Mikolajczyk et al. 'A Comparison of Affine Region Detectors'. In: *International Journal of Computer Vision* 65.1–2 (2005), pp. 43–72.

[Muj+12]   S. Mujagic et al. 'Tactile Conditioning and Movement Analysis of Antennal Sampling Strategies in Honey Bees (Apis Mellifera L.)' In: *Journal of Visualized Experiments* 70 (2012), e50179.

[Mun57]   J. Munkres. 'Algorithms for Assignment and Transportation Problems'. In: *Journal of the Society for Industrial and Applied Mathematics* 5.1 (1957), pp. 32–38.

[Nev+13]   L. Neves et al. 'Iterative Technique for Content-based Image Retrieval using Multiple SVM Ensembles'. In: *Workshop de Visao Computacional*. 2013, pp. 1–6.

[Ney93]   F. Neycenssac. 'Contrast Enhancement Using the Laplacian-of-a-Gaussian Filter'. In: *Graphical Models and Image Processing* 55.6 (1993), pp. 447–463.

[Ngu+09]   M. H. Nguyen et al. 'Weakly Supervised Discriminative Localization and Classification: A Joint Learning Process'. In: *IEEE International Conference on Computer Vision*. 2009, pp. 1925–1932.

[Niy+05]   J. M. Niya et al. '2-step Wavelet-based Edge Detection Using Gabor and Cauchy Directional Wavelets'. In: *International Conference on Advanced Communication Technology*. 2005, pp. 115–120.

[Oka07]   N. Okafor. *Modern Industrial Microbiology and Biotechnology*. Enfield, US: Science Publishers, 2007.

[Orl+07]   N. Orlov et al. 'Computer Vision for Microscopy Applications'. In: *Vision Systems: Segmentation and Pattern Recognition*. Ed. by G. Obinata and A. Dutta. Vienna, Austria: I-Tech, 2007.

[Ots79]   N. Otsu. 'A Threshold Selection Method from Gray-level Histograms'. In: *IEEE Transactions on Systems, Man, and Cybernetics* 9.1 (1979), pp. 62–66.

[PAV15]   D. S. Pham, O. Arandjelovic, and S. Venkatesh. 'Detection of Dynamic Background due to Swaying Movements from Motion Features'. In: *IEEE Transactions on Image Processing* 24.1 (2015), pp. 332–344.

[PCST00]   J. C. Platt, N. Cristianini, and J. Shawe-Taylor. 'Large Margin DAG's for Multiclass Classification'. In: *Advances in Neural Information Processing Systems*. 2000, pp. 547–553.

[PE+14]   A. Perez-Escudero et al. 'IdTracker: Tracking Individuals in a Group by Automatic Identification of Unmarked Animals'. In: *Nature Methods* 11.7 (2014), pp. 743–748.

[PGG14]   I. L. Pepper, C. P. Gerba, and T. J. Gentry. *Environmental Microbiology Third Edition*. San Diego, US: Academic Press, 2014.

[PL11]   M. Pandey and S. Lazebnik. 'Scene Recognition and Weakly Supervised Object Localization with Deformable Part-based Models'. In: *IEEE International Conference on Computer Vision*. 2011, pp. 1307–1314.

[PV02]   M. Pagliai and N. Vignozzi. 'Image Analysis and Microscopic Techniques to Characterize Soil Pore System'. In: *Physical Methods in Agriculture*. Ed. by J. Blahovec and M. Kutilek. New York, US: Springer US, 2002.

[Per+06]   A. G. A. Perera et al. 'Multi-object Tracking Through Simultaneous Long Occlusions and Split-merge Conditions'. In: *IEEE Conference on Computer Vision and Pattern Recognition*. 2006, pp. 666–673.

[Pis+10]   H. Pistori et al. 'Mice and Larvae Tracking Using a Particle Filter with an Auto-adjustable Observation Model'. In: *Pattern Recognition Letters* 31.4 (2010), pp. 337–346.

[Plu+09]   M. D. Plumbley et al. 'Sparse Representations in Audio and Music: From Coding to Source Separation'. In: *Prodeedings of the IEEE* 98.6 (2009), pp. 995–1005.

[Pre70]   J. M. S. Prewitt. 'Object Enhancement and Extraction'. In: *Picture Processing and Psychopictorics*. Academic Press, 1970.

[QS12]   Z. Qin and C. R. Shelton. 'Improving Multi-target Tracking via Social
            Grouping'. In: *IEEE Conference on Computer Vision and Pattern Recognition*.
            2012, pp. 1972–1978.

[RB04]   G. Rangaswami and D. J. Bagyaraj. *Agricultural Microbiology*. India:
            Prentice-Hall of India Pvt. Ltd., 2004.

[RE04]   G. Rabut and J. Ellenberg. 'Automatic Real-time Three-dimensional Cell
            Tracking by Fluorescence Microscopy'. In: *Journal of Microscopy* 216.2
            (2004), pp. 131–137.

[RKG92]   U. Reichl, R. King, and E. D. Gilles. 'Characterization of Pellet Mor-
            phology During Submerged Growth of Streptomyces tendae by Image
            Analysis'. In: *Biotechnology and Bioengineering*, 39 (1992), pp. 164–170.

[RSM11]   R. Rulaningtyas, A. B. Suksmono, and T. L. R. Mengko. 'Automatic Classi-
            fication of Tuberculosis Bacteria Using Neural Network'. In: *International
            Conference on Electrical Engineering and Informatics*. 2011, pp. 1–4.

[RVB84]   A. J. Ringrose-Voase and P. Bullock. 'The Automatic Recognition and
            Measurement of Soil Pore Types by Image Analysis and Computer Pro-
            grams'. In: *Journal of Soil Science* 35.4 (1984), pp. 673–684.

[Rad72]   C. M. Rader. 'Discrete Convolutions via Mersenne Transforms'. In: *IEEE
            Transactions on Computers* 21.12 (1972), pp. 1269–1273.

[Rai+07]   R. Raina et al. 'Self-taught Learning: Transfer Learning from Unlabeled
            Data'. In: *International Conference on Machine Learning*. 2007, pp. 759–766.

[Reh87]   V. Rehder. 'Quantification of the Honeybee's Proboscis Reflex by Elec-
            tromyographic Recordings'. In: *Journal of Insect Physiology* 33.7 (1987),
            pp. 501–503, 505–507.

[Ris+13]   B. Risse et al. 'FIM, a Novel FTIR-based Imaging Method for High
            Throughput Locomotion Analysis'. In: *PLOS One* 8.1 (2013), e53963.

[Rob63]   L. G. Roberts. *Machine Perception of Three-dimensional Solids*. PhD Thesis
            in Massachusetts Institute of Technology. 1963.

[Rod+01]   K. Rodenacker et al. '(Semi-) Automatic Recognition of Microorganisms
            in Water'. In: *International Conference on Image Processing*. 2001, pp. 30–33.

[Rod+04]   A. Rodriguez et al. 'Automatic Analysis of the Content of Cell Biological
            Videos and Database Organization of Their Metadata Descriptors'. In:
            IEEE Transactions on Multimedia 6.1 (2004), pp. 119–128.

[Rod+14]   P. L. Rodrigues et al. 'Automated Image Analysis of Lung Branching Morphogenesis from Microscopic Images of Fetal Rat Explants'. In: *Computational and Mathematical Methods in Medicine* 2014 (2014), 2014. doi:10.1155/2014/820214.

[Ros+05]   B. Rosenhahn et al. *A Silhouette Based Human Motion Tracking System*. Tech. rep. Communication and Information Technology Research Technical Report 164. 2005, pp. 1–33.

[SAT91]   B. H. Smith, C. I. Abramson, and T. R. Tobin. 'Conditional Withholding of Proboscis Extension in Honeybees (Apis Mellifera) during Discriminative Punishment'. In: *Journal of Comparative Psychology* 105.4 (1991), pp. 345–356.

[SB10]   C. J. Solomon and T. P. Breckon. *Fundamentals of Digital Image Processing: A Practical Approach with Examples in Matlab*. West Sussex, UK: Wiley-Blackwell, 2010.

[SF06]   L. Sack and K. Frole. 'Leaf Structural Diversity Is Related to Hydraulic Capacity in Tropical Rain Forest Trees'. In: *Ecology* 87.2 (2006), pp. 483–491.

[SGA95]   H. Salvado, M. P. Gracia, and J. M. Amigo. 'Capability of Ciliated Protozoa as Indicators of Effluent Quality in Activated Sludge Plants'. In: *Water Research* 29.4 (1995), pp. 1041–1050.

[SGU15]   K. Shirahama, M. Grzegorzek, and K. Uehara. 'Weakly Supervised Detection of Video Events Using Hidden Conditional Random Fields'. In: *International Journal on Multimedia Information Retrieval* 4.1 (2015), pp. 17–32.

[SHM11]   C. Smochina, P. Herghelegiu, and V. Manta. *Image Processing Techniques Used in Microscopic Image Segmenation*. Tech. rep. Gheorghe Asachi Technical University of Iasi, 2011.

[SJK09]   Y. Sheikh, O. Javed, and T. Kanade. 'Background Subtraction for Freely Moving Cameras'. In: *IEEE International Conference on Computer Vision*. 2009, pp. 1219–1225.

[SM00]   J. Shi and J. Malik. 'Normalized Cuts and Image Segmentation'. In: *IEEE Transactions on Pattern Analysis and Machine Intelligence* 22.8 (2000), pp. 888–905.

[SN14]   S. Shukla and S. Naganna. 'A Review on k-Means Data Clustering Approach'. In: *International Journal of Information & Computation Technology* 4.17 (2014), pp. 1847–1860.

[SPC13]   D. N. Satange, A. A. Pasarkar, and P. D. Chauhan. 'Microscopic Image Analysis for Plant Tissue Using Image Processing Technique'. In: *International Journal Of Engineering And Computer Science* 2.4 (2013), pp. 1200–1204.

[SR09]   N. Senthilkumaran and R. Rajesh. 'Edge Detection Techniques for Image Segmentation – A Survey of Soft Computing Approaches'. In: *International Journal of Recent Trends in Engineering* 1.2 (2009), pp. 250–254.

[SS04]   M. Sezgin and B. Sankur. 'Survey Over Image Thresholding Techniques and Quantitative Performance Evaluation'. In: *Journal of Electronic Imaging* 13.1 (2004), pp. 146–165.

[SWS05]   C. G. M. Snoek, M. Worring, and A. W. M. Smeulders. 'Early Versus Late Fusion in Semantic Video Analysis'. In: *ACM International Conference on Multimedia*. 2005, pp. 399–402.

[SX11]   P. Siva and T. Xiang. 'Weakly Supervised Object Detector Learning with Model Drift Detection'. In: *IEEE International Conference on Computer Vision*. 2011, pp. 343–350.

[Sak+13]   F. Saki et al. 'Fast Opposite Weight Learning Rules with Application in Breast Cancer Diagnosis'. In: *Computers in Biology and Medicine* 43.1 (2013), pp. 32–41.

[Sar+14]   O. Sarrafzadeh et al. 'Selection of the Best Features for Leukocytes Classification in Blood Smear Microscopic Images'. In: *SPIE 9041, Medical Imaging 2014: Digital Pathology*. 2014, pp. 1–8.

[Sau+03]   S. Sauer et al. 'The Dynamics of Sleep-like Behaviour in Honey Bees'. In: *Journal of Comparative Physiology A* 189.8 (2003), pp. 599–607.

[Sch93]   R. F. Schleif. *Genetics and Molecular Biology Second Edition*. London, UK: The Johns Hopkins University Press, 1993.

[She+13]   M. Shen et al. 'Automatic Framework for Tracking Honeybee's Antennae and Mouthparts from Low Framerate Video'. In: *International Conference on Image Processing*. 2013, pp. 4112–4116.

[She+14]   M. Shen et al. 'Interactive Framework for Insect Tracking with Active Learning'. In: *International Conference on Pattern Recognition*. 2014, pp. 2733–2738.

[She+15a] M. Shen et al. 'Automated Tracking and Analysis of Behaviour in Restrained Insects'. In: *Journal of Neuroscience Methods* 239 (2015), pp. 194–205.

[She+15b] M. Shen et al. 'Interactive Tracking of Insect Posture'. In: *Pattern Recognition* 48.11 (2015), pp. 3560–3571.

[Sib73] R. Sibson. 'SLINK: An Optimally Efficient Algorithm for the Single-link Cluster Method'. In: *The Computer Journal* 16.1 (1973), pp. 30–34.

[Sim+11] M. Simpson et al. *Text- and Content-based Approaches to Image Modality Detection and Retrieval for the ImageCLEF 2010 Medical Retrieval Track.* 2011.

[Sme+00] A. W. M. Smeulders et al. 'Content-based Image Retrieval at the End of the Early Years'. In: *IEEE Transactions on Pattern Analysis and Machine Intelligence* 22.12 (2000), pp. 1349–1380.

[Sob14] I. Sobel. *History and Definition of the Sobel Operator.* On Line. 2014.

[Son+06] Y. Song et al. 'A Novel Miniature Mobile Robot System for Micro Operation Task'. In: *International Conference on Control, Automation, Robotics and Vision.* 2006, pp. 1–6.

[TCY09] L. Tang, J. Chen, and J. Ye. 'On Multiple Kernel Learning with Multiple Labels'. In: *International Joint Conference on Artificial Intelligence.* 2009, pp. 1255–1266.

[TK09] S. Theodoridis and K. Koutroumbas. *Pattern Rcogniton Fourth Edition.* UK: Elsevier, 2009.

[TN84] T. Tsnji and T. Nishikawa. 'Automated Identification of Red Tide Phytoplankton *Prorocentrum* Triestinum in Coastal Areas by Image Analysis'. In: *Journal of the Oceanographical Society of Japan* 40.6 (1984), pp. 425–431.

[TS05] P. Tissainayagama and D. Suterb. 'Object Tracking in Image Sequences Using Point Features'. In: *Pattern Recognition* 38.1 (2005), pp. 105–113.

[TS96] X. Tang and W. K. Stewart. 'Plankton Image Classification Using Novel Parallel-training Learning Vector Quantization Network'. In: *OCEANS'96. MTS/IEEE. Prospects for the 21st Century.* 1996, pp. 1227–1236.

[TSS11] A. Tahmasbi, F. Saki, and S. B. Shokouhi. 'Classification of Benign and Malignant Masses Based on Zernike Moments'. In: *Computers in Biology and Medicine* 41.8 (2011), pp. 726–735.

[TW95] S. Thiel and R. J. Wiltshire. 'The Automated Detection of Cyanobacteria Using Ddigital Image Processing Techniques'. In: *Environment International* 21.2 (1995), pp. 233–236.

[Tan+98]   X. Tang et al. 'Automatic Plankton Image Recognition'. In: *Artificial Intelligence for Biology and Agriculture*. Ed. by S. Panigrahi and K. C. Ting. Netherlands: Kluwer Academic Publishers, 1998.

[Tru+01]   O. Trujillo et al. 'A Machine Vision System Using Immuno-fluorescence Microscopy for Rapid Recognition of Salmonella Typhimurium'. In: *Journal of Rapid Methods & Automation in Microbiology* 9.2 (2001), pp. 115–134.

[Tru+96]   P. Truquet et al. 'Application of a Digital Pattern Recognition System to Dinophysis Acuminata and D-Sacculus Complexes'. In: *Aquatic Living Resources* 9.3 (1996), pp. 273–279.

[USU78]   D. Uhlmann, O. Schlimpeet, and W. Uhlmann. 'Automated Phytoplankton Analysis by a Pattern Recognition Method'. In: *International Review of Hydrobiology* 63.4 (1978), pp. 575–583.

[Uij+13]   J. R. R. Uijlings et al. 'Selective Search for Object Recognition'. In: *International Journal of Computer Vision* 104.2 (2013), pp. 154–171.

[VB11]   C. Venien-Bryan. *Electron Microscopy and Image Analysis I*. On Line. 2011.

[VCL98]   K. Veropoulos, C. Campbell, and G. Learmonth. 'Image Processing and Neural Computing Used in the Diagnosis of Tuberculosis'. In: *IEE Colloquium on Intelligent Methods in Healthcare and Medical Applications*. 1998, pp. 8/1–8/4.

[VCS08]   A. Veeraraghavan, R. Chellappa, and M. Srinivasan. 'Shape-and-Behavior Encoded Tracking of Bee Dances'. In: *IEEE Transactions on Pattern Analysis and Machine Intelligence* 30.3 (2008), pp. 463–476.

[VR11]   C. Vondrick and D. Ramanan. 'Video Annotation and Tracking with Active Learning'. In: *Neural Information Processing Systems*. 2011, pp. 28–36.

[VRP10]   C. Vondrick, D. Ramanan, and D. Patterson. 'Efficiently Scaling up Video Annotation with Crowdsourced Marketplaces'. In: *European Conference on Computer Vision*. 2010, pp. 610–623.

[VS12]   S. R. Vantaram and E. Saber. 'Survey of Contemporary Trends in Colour Image Segmentation'. In: *Journal of Electronic Imaging* 21.4 (2012), pp. 040901–1–040901–28.

[VSC08]   J. Voigts, B. Sakmann, and T. Celikel. 'Unsupervised Whisker Tracking in Unrestrained Behaving Animals'. In: *Journal of Neurophysiology* 100.1 (2008), pp. 504–515.

[Vap98]   V. N. Vapnik. *Statistical Learning Theory*. New York, US: Wiley-Interscience, 1998.

[WLY13]   Y. Wu, J. Lim, and M. H. Yang. 'Online Object Tracking: A Benchmark'. In: *IEEE Conference on Computer Vision and Pattern Recognition*. 2013, pp. 2411–2418.

[WLZ03]   Y. Wen, G. Lu, and X. Zhao. 'Intelligent Control Method on Primitive in Micro-operation Robot'. In: *IEEE International Conference on Robotics, Intelligent Systems and Signal Processing*. 2003, pp. 8–13.

[WMC08]   Q. Wu, F. A. Merchant, and K. R. Castleman. *Microscope Image Processing*. London, UK: Elsevier, 2008.

[WS85]   D. H. Warren and E. R. Strelow. *Electronic Spatial Sensing for the Blind: Contributions from Perception, Rehabilitation, and Computer Vision*. Netherlands: Springer, 1985.

[WSP05]   K. W. Widmer, D. Srikumar, and S. D. Pillai. 'Use of Artificial Neural Networks to Accurately Identify Cryptosporidium Oocyst and Giardia Cyst Images'. In: *Applied and Environmental Microbiology* 71.1 (2005), pp. 80–84.

[WZJ08]   X. Wen, H. Zhang, and Z. Jiang. 'Multiscale Unsupervised Segmentation of SAR Imagery Using the Genetic Algorithm'. In: *Sensors* 8.3 (2008), pp. 1704–1711.

[Whi+12]   M. White et al. 'Convex Multi-view Subspace Learning'. In: *Advances in Neural Information Processing Systems*. 2012, pp. 1682–1690.

[Wil12]   C. Willyard. 'Diagnostics: Playing Detective'. In: *Nature* 491.7425 (2012), S64–S65.

[XTX13]   C. Xu, D. Tao, and C. Xu. 'A Survey on Multi-view Learning'. In: *CoRR* arXiv/1304.5634 (2013).

[XTX14]   C. Xu, D. Tao, and C. Xu. 'Large-margin Multi-view Information Bottleneck'. In: *IEEE Transactions on Pattern Analysis and Machine Intelligence* 36.8 (2014), pp. 1559–1572.

[Xu+10]   Z. Xu et al. 'Simple and Eficient Multiple Kernel Learning by Group Lasso'. In: *International Conference of Machine Learning*. 2010, pp. 1175–1182.

[YKR08]   M. Yang, K. Kpalma, and J. Ronsin. 'A Survey of Shape Feature Extraction Techniques'. In: *Pattern Recognition*. Ed. by P. Yin. IN-TECH, 2008, pp. 43–90.

[YMJ06]   S. Yeom, I. Moon, and B. Javidi. 'Real-time 3-D Sensing, Visualization and Recognition of Dynamic Biological Microorganisms'. In: *Proceedings of IEEE* 94.3 (2006), pp. 550–566.

[YRT14]   J. Yu, Y. Rui, and D. Tao. 'Click Prediction for Web Image Reranking Using Multimodal Sparse Coding'. In: *IEEE Transactions on Image Processing* 23.5 (2014), pp. 2019–2032.

[YWT12]   J. Yu, M. Wang, and D. Tao. 'Semisupervised Multiview Distance Metric Learning for Cartoon Synthesis'. In: *IEEE Transactions on Image Processing* 21.11 (2012), pp. 4636–4648.

[Yan+13] Y. T. Yang et al. 'The Supervised Normalized Cut Method for Detecting, Classifying, and Identifying Special Nuclear Materials'. In: *Informas Journal on Computing* (2013), pp. 45–58.

[Yan+14] C. Yang et al. 'Shape-based Classification of Environmental Microorganisms'. In: *International Conference on Pattern Recognition*. 2014, pp. 3374–3379.

[Yao+12] A. Yao et al. 'Interactive Object Detection'. In: *IEEE Conference on Computer Vision and Pattern Recognition*. 2012, pp. 3242–3249.

[Yin04]   F. Ying. 'Visual Ants Tracking'. PhD thesis. University of Bristol, 2004.

[Yue+09] J. Yuen et al. 'LabelMe Video: Building A Video Database with Human Annotations'. In: *IEEE International Conference on Computer Vision*. 2009, pp. 1451–1458.

[ZL02]   D. Zhang and G. Lu. 'A Comparative Study of Fourier Descriptors for Shape Representation and Retrieval'. In: *Asian Conference on Computer Vision*. 2002, pp. 646–651.

[ZL03]   D. Zhang and G. Lu. 'A Comparative Study of Curvature Scale Space and Fourier Descriptors for Shape-based Image Retrieval'. In: *Journal of Visual Communication and Image Representation* 14.1 (2003), pp. 39–57.

[ZL04]   D. Zhang and G. Lu. 'Review of Shape Representation and Description Techniques'. In: *Pattern Recognition* 37.1 (2004), pp. 1–19.

[ZLN08]   L. Zhang, Y. Li, and R. Nevatia. 'Global Data Association for Multi-object Tracking Using Network Flows'. In: *IEEE Conference on Computer Vision and Pattern Recognition*. 2008, pp. 1–8.

[ZM13]   L. Zhang and L. V. D. Maaten. 'Structure Preserving Object Tracking'. In: *IEEE Conference on Computer Vision and Pattern Recognition*. 2013, pp. 1838–1845.

[ZZW13]   W. Zhang, D. Zhao, and X. Wang. 'Agglomerative Clustering via Maximum Incremental Path Integral'. In: *Pattern Recognition* 46.11 (2013), pp. 3056–3065.

[ZZY12]   K. Zhang, L. Zhang, and M. H. Yang. 'Real-time Compressive Tracking'. In: *European Conference on Computer Vision*. 2012, pp. 864–877.

[Zad98]   L. A. Zadeh. 'Some Reflections on Soft computing, Granular Computing and Their Roles in the Conception, Design and Utilization of Information/Intelligent Systems'. In: *Soft Computing* 2.1 (1998), pp. 23–25.

[Zha+07] C. Zhan et al. 'An Improved Moving Object Detection Algorithm Based on Frame Difference and Edge Detection'. In: *International Conference on Image and Graphics*. 2007, pp. 519–523.

[Zha+12] T. Zhang et al. 'Robust Visual Tracking via Multi-task Sparse Learning'. In: *IEEE Conference on Computer Vision and Pattern Recognition*. 2012, pp. 2042–2049.

[Zou+15] Y. Zou et al. 'Environmental Microbiological Content-based Image Retrieval System Using Internal Structure Histogram'. In: *International Conference on Computer Recognition Systems*. 2015, Accepted for Publication.

List of Publications

Journal Articles in English (3)

2015 C. Li, K. Shirahama, and M. Grzegorzek. 'Environmental Microbiology Aided by Content-based Image Analysis'. In: *Pattern Analysis and Applications*. (2015), Accepted for Publication.

2015 M. Shen, C. Li, W. Huang, P. Szyszka, K. Shirahama, M. Grzegorzek, D. Merhof, and O. Duessen. 'Interactive Tracking of Insect Posture'. In: *Pattern Recognition*. 48.11 (2015), pp. 3560–3571.

2015 C. Li, K. Shirahama, and M. Grzegorzek. 'Application of Content-based Image Analysis to Environmental Microorganism Classification'. In: *Biocybernetics and Biomedical Engineering* 35.1 (2015), pp. 10–21.

Conference Papers in English (8)

2015 Y. Zou, C. Li, Z. Boukhers, T. Jiang, K. Shirahama, and M. Grzegorzek. 'Environmental Microbiological Content-based Image Retrieval System Using Internal Structure Histogram (the best 80 papers)'. In: *International Conference on Computer Recognition Systems*. 2015, Accepted for Publication.

2015 C. Li, K. Shirahama, and M. Grzegorzek. 'Environmental Microorganism Classification Using Sparse Coding and Weakly Supervised Learning (invited paper)'. In: *International Workshop on Environmental Multimedia Retrieval in Conjunction with ACM International Conference on Multimedia Retrieval*. 2015, pp. 9–14.

2014 C. Yang, C. Li, O. Tiebe, K. Shirahama, and M. Grzegorzek. 'The Shape-based Classification of Environmental Microorganisms'. In: *International Conference on Pattern Recognition*. 2014, pp. 3374–3379.

2013 | K. Shirahama, C. Li, M. Grzegorzek, and K. Uehara. 'University of Siegen, Kobe University and Muroran Institute of Technology at TRECVID 2013 Multimedia Event Detection'. In: *TREC Video Retrieval Evaluation (TRECVID) Workshop*. 2013, Page Online.

2013 | C. Li, K. Shirahama, J. Czajkowska, M. Grzegorzek, F. Ma, and B. Zhou. 'A Multi-stage Approach for Automatic Classification of Environmental Microorganisms'. In: *International Conference on Image Processing, Computer Vision, and Pattern Recognition*. 2013, pp. 364–370.

2013 | C. Li, K. Shirahama, M. Grzegorzek, F. Ma, and B. Zhou. 'Classification of Environmental Microorganisms in Microscopic Images Using Shape Features and Support Vector Machines'. In: *IEEE International Conference on Image Processing*. 2013, pp. 2435–2439.

2013 | M. Grzegorzek, C. Li, J. Raskatow, D. Paulus, and N. Vassilieva. 'Texture-based Text Detection in Digital Images with Wavelet Features and Support Vector Machines'. In: *International Conference on Computer Recognition Systems*. 2013, pp. 857–866.

2010 | C. Li, W. Wang, and S. Zhong. 'The Idea of Increasing the Acceptable Levels of Users to AI Products by LCEST (English version). In: *International Conference on Computer Application and System Modeling*. 2010, pp. 4–8.

Books in Chinese (2)

2010 | N. Xu and C. Li. *See the Movies by Oneself*. Shenyang, Liaoning, China: Liaoning Education Press, 2010.

2010 | C. Li and N. Xu. *LC-Equal System Theory*. Shenyang, Liaoning, China: Liaoning Education Press, 2010.

Journal Articles in Chinese (11)

2016 C. Li, Z. Han, X. Huang, M. Shen, Y. Zou, F. Schmidt and T. Jiang. 'A Sobel Method Based Environmental Microorganism Microscopic Image Segmentation System'. In: *Application Research of Computers (Special Issue)*. (2017), Accepted for Publication.

2015 C. Li, W. Yao, Z. Han, Y. Gao, F. Schmidt, T. Jiang, H. Ding, Z. Wang, and M. Shen. 'A Plant Leaf Classification System Using CBIA and WSN'. In: *Application Research of Computers* 32.11 (2015), pp. 3336–3340.

2013 Z. Han, C. Li, and F. Schmidt. 'Multi-class Classification System of Leaves Based on Support Vector Machine and Smart Mobile Devices'. In: *Biotech World* 2013.3 (2013), pp. 173–175.

2012 C. Yang, W. Wang, and C. Li. 'Research on the Problem of Memory Over Flow and Thread Disorder on Tomcat'. In: *Technology and Life* 2012.4 (2012), pp. 108–109.

2010 C. Li. 'The Preliminary Research of the Diagnosis of High School Chemistry Virtual Experiment Operations Using Modeling-based Diagnosis Mentality'. In: *Technology and Life* 2010.20 (2010), pp. 131–132.

2010 C. Li. 'The Research of Organizational Processes Modeling of VR Technology Applied to a Portrait of Criminal Simulation'. In: *Technology and Life* 2010.14 (2010), 111, 125.

2010 C. Li. 'The Idea of Increasing the Acceptable Levels of Users to AI Products by LCEST (Chinese version)'. In: *Technology and Life* 2010.10 (2010), pp. 134–135.

2010 C. Li. 'The Preliminary Research of the Diagnosis of High School Chemistry Virtual Experiment Operations'. In: *Philosophy* 2010.15 (2010), pp. 70.

2010 C. Li and S. Zhong. 'Modeling of Applying Desktop VR System to Client Terminal of Virtual Learning Community'. In: *Journal of Shenyang University* 22.3 (2015), pp. 18–20.

2010 C. Li. 'The Research of the Development of China's Educational Software Market'. In: *Modern Business* 2010.15 (2010), pp. 212–213, 215.

2010 C. Li. 'The Conclusion from the Comparative Analysis of Two Types of Uncertainty Reasoning for the Required Modeling'. In: *Journal of Petroleum Educational Institute of Xinjiang* 2010.1 (2010), pp. 280–281.